Eigenvalues and Eigenvectors

Other Books by Francis D. Hauser

Excel with VBA for Engineers and Mathematicians, 2015. This book includes programs for computing a transfer function, frequency response, root locus, and Dantzig's simplex algorithm.

The Golden Ratio: The Facts and the Myths, 2015. This book looks at the golden ratio from an engineer's viewpoint.

Eigenvalues and Eigenvectors

$$A\bar{x} = \bar{x}\lambda$$

Using

Excel with VBA

(PC and Mac)

Francis D. Hauser

Eigenvalues and Eigenvectors Using Excel with VBA

Copyright © 2016 Francis D. Hauser. All rights reserved.

Excel ® is a registered trademark of Microsoft Corporation.

ISBN-13: 978-1532836329
ISBN-10: 1532836325

Library of Congress Control Number: 2016906953

CreateSpace Independent Publishing Platform, North Charleston, SC

Abstract

The intended audience of this book includes the following:

- Those who know nothing about eigenvalues and eigenvectors
- Those who know something about eigenvalues but nothing about eigenvectors
- Those who know about eigenvalues and eigenvectors but would like working source code that computes the complex eigenvalues and eigenvectors of a general real matrix
- Those who would like to learn about **Excel with VBA**, which is the 21st-century version of the programming language called **BASIC**

The book begins with hand-calculated examples that define *eigenvalues* and *eigenvectors*. These examples show how to compute them, and they demonstrate two important ways to use them.

One way uses them to compute the all-important step response of a dynamic system. The kernels of the response (exponents, sines, cosines) are composed of the eigenvalues. The coefficients that combine the kernels are composed of the eigenvectors. This way uses the eigenvector matrix as a *matrix of rows*.

The other way uses the eigenvector matrix as a *matrix of columns* to form mode shapes. This way exhibits how the responses of the system variables compare with one another at each eigenvalue.

The book then has two chapters that discuss the algorithms used in the VBA programs that compute eigenvalues and eigenvectors.

The next ten chapters define the syntax of the VBA programming language. Every element in the syntax is demonstrated by complete programs that do useful things, such as inverting a complex matrix. Plotting the results and debugging are also demonstrated.

Next comes a chapter that contains complete, checked-out source code for computing eigenvalues and eigenvectors of a general real matrix. This chapter also shows how to input the data and understand the output.

The book concludes with chapters containing example problems showing the above two ways of using eigenvalues and eigenvectors. All of the data to reproduce the example problems are included.

In addition, the book contains numerous methods for checking the results.

Acknowledgment

To Melco Absin and Angie Ilagan-Absin. Thank you both sincerely.

Contents

Introduction and Summary ... 1

1. Examples ... 6
   - Hand calculations defining, computing, and using eigenvalues and eigenvectors

2. A Program to Compute Eigenvalues of a General Real Matrix ... 19
   - Block diagram of a program implementing a QR algorithm

3. A Program to Compute Eigenvectors of a General Real Matrix ... 31
   - Block diagram of a program implementing an exhaustive search

4. The Programming Environment between **Excel** and VBA ... 36
   - Code windows with programs to read from and write to spreadsheets

5. Making Graphs Using VBA ... 41
   - Programs and instructions for X-Y charts and 3-D surface charts

6. VBA Arrays and Data Types ... 50
   - Programs showing the syntax of arrays and data types

7. Solving Linear Algebraic Equations Using Matrices and Dynamic Arrays ... 53
   - Programs showing matrix algebra (inversion and so on)

8. Functions ... 57
   - Programs using **Excel**, VBA, and user-defined functions

9. Looping and Branching and the Operators (Comparison, Logical, and Arithmetic) ... 61
   - The syntax illustrated by code fragments

10. The **Call** Statement ... 70
    - Programs showing the syntax of one program calling another

11. The **GoSub** Statement and Runge-Kutta Numerical Integration ... 84
    - One segment of a program calling another segment

12. The **GoTo** Statement with Sort and Interpolate ... 89
    - A statement causing execution to continue at another statement

13. Debugging   91

- Programs showing menus and windows for troubleshooting

14. VBA Programs to Compute Eigenvalues and Eigenvectors   96

- Instructions for using the programs

15. Applications   105

- Step response of dynamic systems and mode shapes for structural systems

16. More Examples   124

- Situations encountered when computing eigenvectors

17. Factoring a Polynomial Using Eigenvalues   140

- Transforming a polynomial to a matrix

Appendix A: Deriving the Equation for Step Response Using Eigenvectors   141

Appendix B: Deriving the Equation for Step Response with Complex Eigenvalues   143

Appendix C: The Equation for Combining Real and Complex Eigenvalues   144

Appendix D.1: A Program That Implements the LR Algorithm   146

Appendix D.2: A Program That Implements the QR Algorithm   147

Appendix D.3: Subprogram eig Code   148

Appendix D.4: Subprogram eigvec Code   151

Appendix E: Code Fragments for Program GoSub_main in chapter 11   155

Appendix F: Case 3, chapter 15 Eigenvector and Lamda Matrices   157

Index: The VBA Language Elements   158

Introduction and Summary

The eigenvalues and eigenvectors of matrix A are computed from the following equation:

$$A\bar{x} = \bar{x}\lambda, \qquad \text{Equation A}$$

where
- A is a square matrix of real numbers;
- λ is a number that may be complex and is called an eigenvalue of A; and
- $\bar{x}$ is a column matrix and is the eigenvector associated with λ.

This is a strange-looking equation. It has had a profound impact on science, engineering, and mathematics. There are countless references about it. Wilkinson's book is an important one.[1]

Equation A has two unknowns and is nonlinear. The following method to solve it is ingenious as well as obvious. I show it using an example.

Given $A = \begin{bmatrix} 0 & 1 \\ -2 & -3 \end{bmatrix}$. Define $\bar{x} = \begin{bmatrix} x_1 \\ x_2 \end{bmatrix}$. Equation A becomes: $\begin{bmatrix} 0 & 1 \\ -2 & -3 \end{bmatrix}\begin{bmatrix} x_1 \\ x_2 \end{bmatrix} = \begin{bmatrix} x_1 \\ x_2 \end{bmatrix}\lambda$.

Rearranging: $\begin{bmatrix} -\lambda & 1 \\ -2 & -3-\lambda \end{bmatrix}\begin{bmatrix} x_1 \\ x_2 \end{bmatrix} = 0$. We will call these equations the *Base Equations*.

If the matrix $\begin{bmatrix} -\lambda & 1 \\ -2 & -3-\lambda \end{bmatrix}$ has an inverse, then $\begin{bmatrix} x_1 \\ x_2 \end{bmatrix} = 0$. We don't want that solution. To prevent this, we want the determinant of $\begin{bmatrix} -\lambda & 1 \\ -2 & -3-\lambda \end{bmatrix}$ to equal zero.

Computing this determinant: $\begin{vmatrix} -\lambda & 1 \\ -2 & -3-\lambda \end{vmatrix} = \lambda^2 + 3\lambda + 2 = (\lambda+1)(\lambda+2) = 0$. It equals zero for two values of λ, namely, $\lambda = -1$ and $\lambda = -2$. Therefore, A has two eigenvalues. We can now compute $\bar{x}$.

- When $\lambda = -1$, the *Base Equations* become: $\begin{bmatrix} 1 & 1 \\ -2 & -2 \end{bmatrix}\begin{bmatrix} x_1 \\ x_2 \end{bmatrix} = 0$. In equation form, they are

(E1) $x_1 + x_2 = 0$
(E2) $-2x_1 - 2x_2 = 0$. There are an infinite number of solutions. We can arbitrarily pick the value of one of the variables and solve for the other using either equation. Let's pick $x_2 = 1$, and then compute $x_1 = -1$. Therefore, a suitable eigenvector for $\lambda = -1$ is $\bar{x} = \begin{bmatrix} -1 & 1 \end{bmatrix}^T$. It is referred to as *suitable* because the value of x_2 was a choice.

---

[1] J. H. Wilkinson, *The Algebraic Eigenvalue Problem* (Oxford: Oxford University Press, 1965).

- When $\lambda = -2$, the Base Equations become: $\begin{bmatrix} 2 & 1 \\ -2 & -1 \end{bmatrix} \begin{bmatrix} x_1 \\ x_2 \end{bmatrix} = 0$. In equation form, they are

(E1) $2x_1 + x_2 = 0$
(E2) $-2x_1 - x_2 = 0$. Again there are an infinite number of solutions. Again let's pick $x_2 = 1$. We compute $x_1 = -0.5$. A suitable eigenvector for $\lambda = -2$ is $\bar{x} = \begin{bmatrix} -0.5 & 1 \end{bmatrix}^T$.

Let's combine these eigenvectors into a single matrix: $V = \begin{bmatrix} -1 & -0.5 \\ 1 & 1 \end{bmatrix}$.

Let's combine the eigenvalues into a single matrix: $L = \begin{bmatrix} -1 & 0 \\ 0 & -2 \end{bmatrix}$.

We can now write Equation A in matrix form:

$$AV = VL. \qquad \text{Equation B}$$

Substituting values, we get:

$$\begin{bmatrix} 0 & 1 \\ -2 & -3 \end{bmatrix} \begin{bmatrix} -1 & -0.5 \\ 1 & 1 \end{bmatrix} = \begin{bmatrix} -1 & -0.5 \\ 1 & 1 \end{bmatrix} \begin{bmatrix} -1 & 0 \\ 0 & -2 \end{bmatrix}.$$ Equation B checks out.

From Equation B, we see that $V^{-1}AV = L = \begin{bmatrix} -1 & 0 \\ 0 & -2 \end{bmatrix} = \begin{bmatrix} \lambda_1 & 0 \\ 0 & \lambda_2 \end{bmatrix}$. The eigenvectors of A have diagonalized it. We can make use of this fact to find the time response of a dynamic system. The following notation will be used: $\dot{z} = \dfrac{d}{dt} z$.

The A matrix is used in the following system:

$$\dot{Z}_1 = Z_2$$
$$\dot{Z}_2 = -2Z_1 - 3Z_2 + 2U$$

where U is the input and Z_1 is the output, which will be called Y. In matrix form:

$$\begin{bmatrix} \dot{Z}_1 \\ \dot{Z}_2 \end{bmatrix} = \begin{bmatrix} 0 & 1 \\ -2 & -3 \end{bmatrix} \begin{bmatrix} Z_1 \\ Z_2 \end{bmatrix} + \begin{bmatrix} 0 \\ 2 \end{bmatrix} U \quad \text{and} \quad Y = \begin{bmatrix} 1 & 0 \end{bmatrix} \begin{bmatrix} Z_1 \\ Z_2 \end{bmatrix}.$$

Let's name the matrices: $A = \begin{bmatrix} 0 & 1 \\ -2 & -3 \end{bmatrix}$, $B = \begin{bmatrix} 0 \\ 2 \end{bmatrix}$, and $C = \begin{bmatrix} 1 & 0 \end{bmatrix}$. The equations become:

$$\begin{bmatrix} \dot{Z}_1 \\ \dot{Z}_2 \end{bmatrix} = A \begin{bmatrix} Z_1 \\ Z_2 \end{bmatrix} + B*U \quad \text{and} \quad Y = C \begin{bmatrix} Z_1 \\ Z_2 \end{bmatrix}. \qquad \text{The System Equations}$$

Introduction and Summary

We would not try to solve these equations by hand. We would have to use numerical integration. Let's see how eigenvectors can be used. Let's use the inverse of the eigenvector matrix to change the Z coordinates to another set of coordinates called q:

$$\begin{bmatrix} q_1 \\ q_2 \end{bmatrix} = V^{-1} \begin{bmatrix} Z_1 \\ Z_2 \end{bmatrix} \text{ where } V^{-1} = \begin{bmatrix} -2 & -1 \\ 2 & 2 \end{bmatrix}.$$

With this transformation:
$$\begin{bmatrix} Z_1 \\ Z_2 \end{bmatrix} = V \begin{bmatrix} q_1 \\ q_2 \end{bmatrix} \text{ and } \begin{bmatrix} \dot{Z}_1 \\ \dot{Z}_2 \end{bmatrix} = V \begin{bmatrix} \dot{q}_1 \\ \dot{q}_2 \end{bmatrix}.$$

The System Equations become:
$$V \begin{bmatrix} \dot{q}_1 \\ \dot{q}_2 \end{bmatrix} = A V \begin{bmatrix} q_1 \\ q_2 \end{bmatrix} + B U.$$

Continuing:
$$\begin{bmatrix} \dot{q}_1 \\ \dot{q}_2 \end{bmatrix} = V^{-1} A V \begin{bmatrix} q_1 \\ q_2 \end{bmatrix} + V^{-1} B U.$$

We have shown that $V^{-1} A V = \begin{bmatrix} \lambda_1 & 0 \\ 0 & \lambda_2 \end{bmatrix}$. We can compute: $V^{-1} B = \begin{bmatrix} -2 & -1 \\ 2 & 2 \end{bmatrix} \begin{bmatrix} 0 \\ 2 \end{bmatrix} = \begin{bmatrix} -2 \\ 4 \end{bmatrix} = \begin{bmatrix} vB_1 \\ vB_2 \end{bmatrix}.$

Finally:
$$\begin{bmatrix} \dot{q}_1 \\ \dot{q}_2 \end{bmatrix} = \begin{bmatrix} \lambda_1 & 0 \\ 0 & \lambda_2 \end{bmatrix} \begin{bmatrix} q_1 \\ q_2 \end{bmatrix} + \begin{bmatrix} vB_1 \\ vB_2 \end{bmatrix} U.$$

These are independent first-order differential equations:

$$\dot{q}_1 = \lambda_1 q_1 + vB_1 U \quad \text{and} \quad \dot{q}_2 = \lambda_2 q_2 + vB_2 U.$$

This type of equation was the first differential equation I learned to solve in school. I learned how to derive its response when U = 1 (the derivation is in appendix A):

$$q_1 = \frac{vB_1}{\lambda_1}(e^{\lambda_1 t} - 1) \quad \text{and} \quad q_2 = \frac{vB_2}{\lambda_2}(e^{\lambda_2 t} - 1).$$

Earlier we said that: $Y = C \begin{bmatrix} Z_1 \\ Z_2 \end{bmatrix}$, which becomes $Y = C V \begin{bmatrix} q_1 \\ q_2 \end{bmatrix}.$

Now: $C V = \begin{bmatrix} 1 & 0 \end{bmatrix} \begin{bmatrix} -1 & -0.5 \\ 1 & 1 \end{bmatrix} = \begin{bmatrix} -1 & -0.5 \end{bmatrix} = \begin{bmatrix} Cv_1 & Cv_2 \end{bmatrix}$. Hence, $Y = Cv_1 q_1 + Cv_2 q_2$.

Therefore:
$$Y = \frac{Cv_1 vB_1}{\lambda_1}(e^{\lambda_1 t} - 1) + \frac{Cv_2 vB_2}{\lambda_2}(e^{\lambda_2 t} - 1). \qquad \text{Equation X}$$

Plugging numbers:
$$Y = -2(e^{-t} - 1) + (e^{-2t} - 1). \qquad \text{Equation Y}$$

The following figure shows a plot of Y. It also shows the output that resulted from numerically integrating the **System Equations**. (The integrator is a *fourth-order Runge-Kutta*, which is described in chapter 11.) The plot shows that the eigenvalues and eigenvectors are accurate.

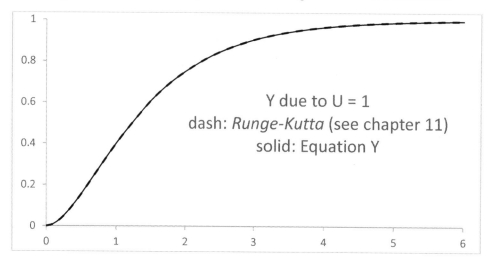

For this example, there were no rounding errors in the computations for the eigenvalues and eigenvectors. Hence, **Equation Y** is exact. The results from the numerical integrator are always approximate.

The role of the eigenvalues and the eigenvectors is clearly shown in **Equation Y**. The response is characterized by the eigenvalues. For this case, the response is characterized as *exponential*. In fact, the equation used to compute eigenvalues ($|A - \lambda I| = 0$) is called the *characteristic* equation. The eigenvectors determine how each eigenvalue contributes to the total response.

For complex eigenvalues, the procedure is identical down to **Equation X**. The coefficients and the eigenvalues in **Equation X** will be complex conjugates. Since $e^{jx} = \cos x + j \sin x$, both terms in **Equation X** combine into sine and cosine terms. This will be shown in example 1 of chapter 1.

Book Summary

Chapter 1 contains hand-calculated examples that further define *eigenvalues* and *eigenvectors* and show two important ways to use them.

One way uses them to compute the all-important step response of a dynamic system as shown in **Equation Y** above. The kernels of the response (exponents, sines, cosines) are composed of the eigenvalues. The coefficients that combine the kernels are composed of the eigenvectors. This way uses the eigenvector matrix as a *matrix of rows*.

The other way uses the eigenvector matrix as a *matrix of columns* to form mode shapes. This way exhibits how the responses of the system variables compare with one another at each eigenvalue. This will be shown in example 4 of chapter 1.

The strategy used to compute eigenvalues and eigenvectors is discussed in chapters 2 and 3.

Chapters 4–13 define the syntax of the VBA programming language. Every element in the syntax is demonstrated by complete programs that do useful things, such as inverting a complex matrix. Plotting the results and debugging are also demonstrated.

Chapter 14 contains complete, checked-out source code for computing eigenvalues and eigenvectors of a general real matrix. This chapter also shows how to input the data and understand the output.

Chapters 15 and 16 contain example problems showing the above two ways of using eigenvalues and eigenvectors. All of the data to reproduce the example problems are included.

Factoring polynomials using eigenvalues is the topic of chapter 17.

Chapter 1: Examples

In the introduction, we used an example to show how eigenvalues and eigenvectors of a matrix are defined, computed, and can be used. In this chapter, I provide some more examples. For continuity, I repeat the defining equation:

$$A\bar{x} = \bar{x}\lambda.$$ Equation A

The literature refers to this equation as the *Eigenvalue Problem*.

Example 1: Complex Eigenvalues

Given $A = \begin{bmatrix} 0 & 1 \\ -2 & -0.5 \end{bmatrix}$. Define $\bar{x} = \begin{bmatrix} x_1 \\ x_2 \end{bmatrix}$. Substituting into Equation A, $\begin{bmatrix} 0 & 1 \\ -2 & -0.5 \end{bmatrix}\begin{bmatrix} x_1 \\ x_2 \end{bmatrix} = \begin{bmatrix} x_1 \\ x_2 \end{bmatrix}\lambda$.

Rewriting: $\begin{bmatrix} -\lambda & 1 \\ -2 & -0.5-\lambda \end{bmatrix}\begin{bmatrix} x_1 \\ x_2 \end{bmatrix} = 0$. We will call these equations the Base Equations.

For $\begin{bmatrix} x_1 \\ x_2 \end{bmatrix} \neq 0$, we need $\begin{vmatrix} -\lambda & 1 \\ -2 & -0.5-\lambda \end{vmatrix} = 0$. The polynomial of this determinant is $\lambda^2 + 0.5\lambda + 2$.

Its roots are $\lambda = -0.25 + j1.39$ and $\lambda = -0.25 - j1.39$. Therefore, A has two eigenvalues. We can now compute $\bar{x}$ from the Base Equations.

- When $\lambda = -0.25 + j1.39$, the Base Equations become: $\begin{bmatrix} 0.25 - j1.39 & 1 \\ -2 & -0.25 - j1.39 \end{bmatrix}\begin{bmatrix} x_1 \\ x_2 \end{bmatrix} = 0$.

There are an infinite number of solutions. We can arbitrarily pick the value of one of the variables. If there are no round-off errors, we can solve for the other variable using either one of the Base Equations. Let's pick $x_2 = 1$. We compute $x_1 = -0.125 - j0.695$. Therefore, a suitable eigenvector for $\lambda = -0.25 + j1.39$ is $\bar{x} = \begin{bmatrix} -0.125 - j0.695 & 1 \end{bmatrix}^T$.

- When $\lambda = -0.25 - j1.39$, the Base Equations become: $\begin{bmatrix} 0.25 + j1.39 & 1 \\ -2 & -0.25 + j1.39 \end{bmatrix}\begin{bmatrix} x_1 \\ x_2 \end{bmatrix} = 0$

Again there are an infinite number of solutions. Again let's pick $x_2 = 1$. We compute $x_1 = -0.125 + j0.695$. Our chosen eigenvector for $\lambda = -0.25 - j1.39$ is $\bar{x} = \begin{bmatrix} -0.125 + j0.695 & 1 \end{bmatrix}^T$

Note: When the eigenvalues are complex conjugates, so are the eigenvectors.

Combining these eigenvectors into a single matrix: $V = \begin{bmatrix} -0.125 - j0.695 & -0.125 + j0.695 \\ 1 & 1 \end{bmatrix}$.

Combining the eigenvalues into a single matrix: $L = \begin{bmatrix} -0.25 + j1.39 & 0 \\ 0 & -0.25 - j1.39 \end{bmatrix}$.

We can now write **Equation A** in matrix form:

$$AV = VL. \qquad \text{Equation B}$$

When we substitute values, **Equation B** checks out.

From **Equation B**, we see that $V^{-1}AV = L = \begin{bmatrix} \lambda_1 & 0 \\ 0 & \lambda_2 \end{bmatrix}$. The eigenvectors of A have diagonalized it. We will use this fact to find the time response of a dynamic system. In the following application, this notation will be used: $\dot{Z} = \dfrac{d}{dt} Z$.

The task is to determine the step response of the following system, which has the same A matrix:

$$\begin{aligned} \dot{Z}_1 &= Z_2 \\ \dot{Z}_2 &= -2Z_1 - 0.5Z_2 + 2U \end{aligned} \qquad \text{The System Equations}$$

where U is the input and Z_1 is the output, which will be called Y. These equations form a system since all of them must be solved together.

In matrix form: $\begin{bmatrix} \dot{Z}_1 \\ \dot{Z}_2 \end{bmatrix} = \begin{bmatrix} 0 & 1 \\ -2 & -0.5 \end{bmatrix} \begin{bmatrix} Z_1 \\ Z_2 \end{bmatrix} + \begin{bmatrix} 0 \\ 2 \end{bmatrix} U$ and $Y = \begin{bmatrix} 1 & 0 \end{bmatrix} \begin{bmatrix} Z_1 \\ Z_2 \end{bmatrix}$.

Let's name the matrices: $A = \begin{bmatrix} 0 & 1 \\ -2 & -0.5 \end{bmatrix}$, $B = \begin{bmatrix} 0 \\ 2 \end{bmatrix}$, and $C = \begin{bmatrix} 1 & 0 \end{bmatrix}$. We can now write the

equations as: $\begin{bmatrix} \dot{Z}_1 \\ \dot{Z}_2 \end{bmatrix} = A \begin{bmatrix} Z_1 \\ Z_2 \end{bmatrix} + B*U$ and $Y = C \begin{bmatrix} Z_1 \\ Z_2 \end{bmatrix}$ **The System Equations**

The inverse of the eigenvector matrix will be used to change the Z coordinates to another set of coordinates called q.

$$\begin{bmatrix} q_1 \\ q_2 \end{bmatrix} = V^{-1} \begin{bmatrix} Z_1 \\ Z_2 \end{bmatrix} \text{ where } V^{-1} = \begin{bmatrix} j0.718 & 0.5 + j0.09 \\ -j0.718 & 0.5 - j0.09 \end{bmatrix}.$$

With this transformation: $\begin{bmatrix} Z_1 \\ Z_2 \end{bmatrix} = V \begin{bmatrix} q_1 \\ q_2 \end{bmatrix}$ and $\begin{bmatrix} \dot{Z}_1 \\ \dot{Z}_2 \end{bmatrix} = V \begin{bmatrix} \dot{q}_1 \\ \dot{q}_2 \end{bmatrix}$.

The System Equations become:
$$V \begin{bmatrix} \dot{q}_1 \\ \dot{q}_2 \end{bmatrix} = A V \begin{bmatrix} q_1 \\ q_2 \end{bmatrix} + B U.$$

Continuing:
$$\begin{bmatrix} \dot{q}_1 \\ \dot{q}_2 \end{bmatrix} = V^{-1} A V \begin{bmatrix} q_1 \\ q_2 \end{bmatrix} + V^{-1} B U.$$

We have shown that $V^{-1} A V = \begin{bmatrix} \lambda_1 & 0 \\ 0 & \lambda_2 \end{bmatrix}$. We need to compute: $V^{-1} B = \begin{bmatrix} vB_1 \\ vB_2 \end{bmatrix} = \begin{bmatrix} 1 + j0.18 \\ 1 - j0.18 \end{bmatrix}$.

Finally:
$$\begin{bmatrix} \dot{q}_1 \\ \dot{q}_2 \end{bmatrix} = \begin{bmatrix} \lambda_1 & 0 \\ 0 & \lambda_2 \end{bmatrix} \begin{bmatrix} q_1 \\ q_2 \end{bmatrix} + \begin{bmatrix} vB_1 \\ vB_2 \end{bmatrix} U.$$

These are independent first-order differential equations:

$$\dot{q}_1 = \lambda_1 q_1 + vB_1 U \quad \text{and} \quad \dot{q}_2 = \lambda_2 q_2 + vB_2 U.$$

I noted in the introduction that this type of equation was the first differential equation I learned to solve in school. I learned how to derive its response when $U = 1$ (see appendix A for the derivation):

$$q_1 = \frac{vB_1}{\lambda_1}(e^{\lambda_1 t} - 1) \quad \text{and} \quad q_2 = \frac{vB_2}{\lambda_2}(e^{\lambda_2 t} - 1).$$

Earlier we said that: $Y = C \begin{bmatrix} Z_1 \\ Z_2 \end{bmatrix}$, which becomes $Y = C V \begin{bmatrix} q_1 \\ q_2 \end{bmatrix}$.

Now, $C V = \begin{bmatrix} Cv_1 & Cv_2 \end{bmatrix} = \begin{bmatrix} -0.125 - j0.695 & -0.125 + j0.695 \end{bmatrix}$. Hence $Y = Cv_1 q_1 + Cv_2 q_2$

Therefore:
$$Y = \frac{Cv_1 vB_1}{\lambda_1}(e^{\lambda_1 t} - 1) + \frac{Cv_2 vB_2}{\lambda_2}(e^{\lambda_2 t} - 1). \qquad \text{Equation X}$$

Since both the eigenvalues and eigenvectors are conjugates, it is expected that the coefficients in Equation X are conjugates. In fact, they have to be if Y is the response of a real system. Hence Equation X can be written as:

$$Y = (a_c + jb_c)(e^{(\lambda_R + j\lambda_I)t} - 1) + (a_c - jb_c)(e^{(\lambda_R - j\lambda_I)t} - 1). \qquad \text{Equation X}$$

In appendix B, it is shown that Equation X can be further written as:

$$Y = 2\{e^{\lambda_R t}[a_c \cos(\lambda_I t) - b_c \sin(\lambda_I t)] - a_c\}. \qquad \text{Equation X}$$

Plugging numbers: $Y = 2\{e^{-0.25t}[-0.5\cos(1.39t) - 0.09\sin(1.39t)] + 0.5\}$. \qquad Equation Y

The following figure shows a plot of Y versus time. It also shows the **output** that resulted from numerically integrating **the System Equations** using the *fourth-order Runge-Kutta* method, which is described in chapter 11. The responses are virtually identical.

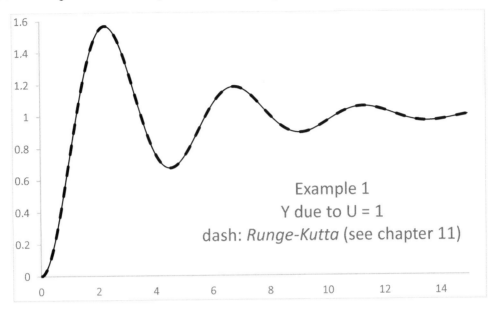

Example 1
Y due to U = 1
dash: *Runge-Kutta* (see chapter 11)

The role of the eigenvalues and the eigenvectors is clearly shown in **Equation Y**. The eigenvalues are computed from the *characteristic* equation $|A - \lambda I| = 0$. They characterize the response. For this case, the response is denoted as *exponentially decaying sinusoid*. The eigenvectors determine how each eigenvalue contributes to the total response.

Examples

Example 2: Real and Complex Eigenvalues

We can now develop a step-by-step approach for computing eigenvalues and eigenvectors.

Given:
$$A = \begin{bmatrix} 0 & 1 & 0 \\ -2 & -0.5 & 2 \\ 0 & 0 & -10 \end{bmatrix}.$$

Step 1: Form the **Base Equations**.

$$\begin{bmatrix} -\lambda & 1 & 0 \\ -2 & -0.5-\lambda & 2 \\ 0 & 0 & -10-\lambda \end{bmatrix} \begin{bmatrix} x_1 \\ x_2 \\ x_3 \end{bmatrix} = 0. \qquad \text{The Base Equations}$$

Step 2: Solve the *characteristic* equation.

$$\begin{vmatrix} -\lambda & 1 & 0 \\ -2 & -0.5-\lambda & 2 \\ 0 & 0 & -10-\lambda \end{vmatrix} = 0.$$

To evaluate this determinant, expand it in minors about the last row. The resulting polynomial is:

$$(-10 - \lambda)(\lambda^2 + 0.5\lambda + 2).$$

The roots of the polynomial are the eigenvalues of A:

$$\lambda_1 = -10, \lambda_2 = -0.25 + j1.39, \text{ and } \lambda_3 = -0.25 - j1.39.$$

Step 3: For each of these eigenvalues, compute its eigenvector from the **Base Equations**.

- For $\lambda_1 = -10$, the Base Equations are: $\begin{bmatrix} 10 & 1 & 0 \\ -2 & 9.5 & 2 \\ 0 & 0 & 0 \end{bmatrix} \begin{bmatrix} x_1 \\ x_2 \\ x_3 \end{bmatrix} = 0$. Pick $x_3 = 1$. Compute

$x_1 = 0.0206$ and $x_2 = -0.206$. Hence, for this eigenvalue, a suitable eigenvector is

$$\bar{x} = \begin{bmatrix} 0.0206 & -0.206 & 1 \end{bmatrix}^T.$$

- For $\lambda_2 = -0.25 + j1.39$, the Base Equations are:

$$\begin{bmatrix} 0.25-j1.39 & 1 & 0 \\ -2 & -0.25-j1.39 & 2 \\ 0 & 0 & -9.75-j1.39 \end{bmatrix} \begin{bmatrix} x_1 \\ x_2 \\ x_3 \end{bmatrix} = 0.$$

The last equation will result in $x_3 = 0$. So we will then pick $x_2 = 1$. With this, we compute that $x_1 = -0.125 - j\,0.696$ and $x_3 = 0$. Hence, a suitable eigenvector is:

$$\bar{x} = \begin{bmatrix} -0.125 - j0.696 & 1 & 0 \end{bmatrix}^T.$$

- For $\lambda_3 = -0.25 - j1.39$, since λ_2 and λ_3 are conjugates, their eigenvectors are also conjugates. Hence, the eigenvector is $\bar{x} = \begin{bmatrix} -0.125 + j0.696 & 1 & 0 \end{bmatrix}^T$.

Step 4: Combine the eigenvectors into a single matrix.

$$V = \begin{bmatrix} 0.0206 & -0.125-j0.696 & -0.125+j0.696 \\ -0.206 & 1 & 1 \\ 1 & 0 & 0 \end{bmatrix}.$$

Compute:
$$V^{-1} = \begin{bmatrix} 0 & 0 & 1 \\ j0.718 & 0.5+j0.09 & 0.103+j0.0037 \\ -j0.718 & 0.5-j0.09 & 0.103-j0.0037 \end{bmatrix}.$$

Finally:
$$V^{-1}AV = L = \begin{bmatrix} -10 & 0 & 0 \\ 0 & -0.25+j1.39 & 0 \\ 0 & 0 & -0.25-j1.39 \end{bmatrix}.$$

Step 5: Use the above eigenvalues and eigenvectors to compute the step response of a system that has the same A matrix.

$$\begin{aligned} \dot{Z}_1 &= Z_2 \\ \dot{Z}_2 &= 2*(Z_3 - Z_1) - 0.5 Z_2 \qquad \text{The System Equations} \\ \dot{Z}_3 &= 10*(U - Z_3) \end{aligned}$$

where U is the input and Z_1 is the output (Y).

In matrix form:
$$\begin{bmatrix} \dot{Z}_1 \\ \dot{Z}_2 \\ \dot{Z}_3 \end{bmatrix} = \begin{bmatrix} 0 & 1 & 0 \\ -2 & -0.5 & 2 \\ 0 & 0 & -10 \end{bmatrix} \begin{bmatrix} Z_1 \\ Z_2 \\ Z_3 \end{bmatrix} + \begin{bmatrix} 0 \\ 0 \\ 10 \end{bmatrix} U.$$

The B and C matrices are: $B = \begin{bmatrix} 0 & 0 & 10 \end{bmatrix}^T$ and $C = \begin{bmatrix} 1 & 0 & 0 \end{bmatrix}$.

Transform this system into the following set of first-order differential equations:

$$\dot{q}_1 = \lambda_1 q_1 + vB_1 U, \quad \dot{q}_2 = \lambda_2 q_2 + vB_2 U, \quad \dot{q}_3 = \lambda_3 q_3 + vB_3 U, \quad \text{and} \quad Y = CV q.$$

The q coordinates are called modal coordinates.

To use these equations, we need to compute the following matrices:

$$V^{-1}B = \begin{bmatrix} vB_1 \\ vB_2 \\ vB_3 \end{bmatrix} = \begin{bmatrix} 10 \\ 1.03+j0.037 \\ 1.03-j0.037 \end{bmatrix} \text{ and } [CV] = \begin{bmatrix} Cv_1 \\ Cv_2 \\ Cv_3 \end{bmatrix}^T = \begin{bmatrix} 0.02 \\ -0.125-j0.696 \\ -0.125+j0.696 \end{bmatrix}^T.$$

We now have the data for the output equation when U = 1.

$$Y = \frac{Cv_1 vB_1}{\lambda_1}(e^{\lambda_1 t} - 1) + \frac{Cv_2 vB_2}{\lambda_2}(e^{\lambda_2 t} - 1) + \frac{Cv_3 vB_3}{\lambda_3}(e^{\lambda_3 t} - 1).$$

In this equation: $\frac{Cv_2 vB_2}{\lambda_2} = a_c + jb_c$, $\lambda_2 = \lambda_R + j\lambda_I$, $\frac{Cv_3 vB_3}{\lambda_3} = a_c - jb_c$, and $\lambda_3 = \lambda_R - j\lambda_I$.

The output equation can now be written as:

$$Y = \frac{Cv_1 vB_1}{\lambda_1}(e^{\lambda_1 t} - 1) + 2\{e^{\lambda_R t}[a_c \cos(\lambda_I t) - b_c \sin(\lambda_I t)] - a_c\}.$$

Data for this equation are: $\frac{Cv_1 vB_1}{\lambda_1} = -0.02$, $\lambda_1 = -10$, $\lambda_R = -0.25$, $\lambda_I = 1.39$, $a_c = -0.49$, and $b_c = 0.16$.

Plugging numbers: $Y = -0.02(e^{-10t} - 1) + 2\{e^{-0.25t}[-0.49\cos(1.39t) - 0.16\sin(1.39t)] + 0.49\}.$

The following figure shows the plot of Y versus time. It also shows the output that resulted from numerically integrating the system equations using the fourth-order Runge-Kutta described in chapter 11. The responses are virtually identical.

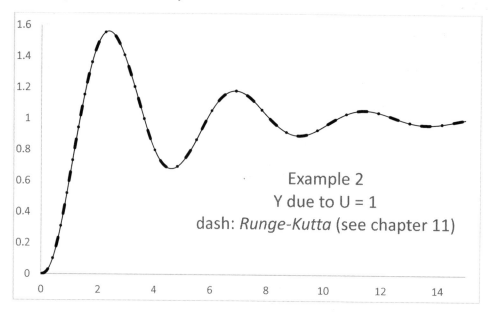

Example 2
Y due to U = 1
dash: *Runge-Kutta* (see chapter 11)

Example 3: Round-Off Errors

For systems higher than second order, eigenvalues are computed iteratively. Round-off errors always occur. In this example, we study this.

Given: $A = \begin{bmatrix} 0 & 1 \\ -2 & -3 \end{bmatrix}$. This matrix was used in the introduction. Its inherent eigenvalues are $\lambda_1 = -1$ and $\lambda_2 = -2$. Suppose that an eigenvalue program had computed $\lambda_1 = -1.05$ and $\lambda_2 = -2.05$.

We begin by forming the **Base Equations**: $\begin{bmatrix} -\lambda & 1 \\ -2 & -3-\lambda \end{bmatrix} \begin{bmatrix} x_1 \\ x_2 \end{bmatrix} = 0$.

In equation form:
$$\begin{aligned}(E1)\quad &-\lambda x_1 + x_2 = 0 \\ (E2)\quad &-2x_1 + (-3-\lambda)x_2 = 0\end{aligned}$$

- The eigenvector for $\lambda_1 = -1.05$ is computed as follows. Pick $x_2 = 1$.
    - Equation (E1) becomes $[1.05x_1 + 1 = 0]$, which yields $x_1 = -0.952$.
    - Equation (E2) becomes $[-2x_1 - 1.95 = 0]$, which yields $x_1 = -0.975$.

Because of the error in the eigenvalue, the **Base Equations** contradict. How can we proceed? Arbitrarily, we will pick the value that results from (E2). The eigenvector will then be

$$\overline{x} = \begin{bmatrix} -.975 & 1 \end{bmatrix}^T.$$

- The eigenvector for $\lambda_2 = -2.05$ is similarly computed. Pick $x_2 = 1$.
    - Equation (E1) becomes $[2.05x_1 + 1 = 0]$, which yields $x_1 = -0.488$.
    - Equation (E2) becomes $[-2x_1 - 0.95 = 0]$, which yields $x_1 = -0.475$.

Again, we will use the result from (E2). The eigenvector will then be $\overline{x} = \begin{bmatrix} -.475 & 1 \end{bmatrix}^T$.

We now form the combined matrices as shown in the following table.

Combined Matrices		
$V = \begin{bmatrix} -0.975 & -0.475 \\ 1 & 1 \end{bmatrix}$	$V^{-1} = \begin{bmatrix} -2 & -0.95 \\ 2 & 1.95 \end{bmatrix}$	$L = V^{-1}AV = \begin{bmatrix} -1.0025 & -0.0525 \\ -0.0475 & -1.9975 \end{bmatrix}$

The round-off errors show up as nonzero off-diagonal elements in the L matrix. But the diagonal elements are pretty close to the actual eigenvalues of the A matrix. This is logical since the defining equation $A\overline{x} = \overline{x}\lambda$ seeks to compute a diagonal matrix with the eigenvalues of A on its diagonal.

We will proceed computing a step response using the diagonal elements of L as if they were the eigenvalues of the A matrix. The system we will use is shown in the following table.

The System Equations			
$\dot{Z}_1 = Z_2$ $\dot{Z}_2 = 2(U - Z_1) - 3Z_2$ $Y = Z_1$	$A = \begin{bmatrix} 0 & 1 \\ -2 & -3 \end{bmatrix}$	$B = \begin{bmatrix} 0 \\ 2 \end{bmatrix}$	$C = \begin{bmatrix} 1 \\ 0 \end{bmatrix}^T$

For this example, the step-response equation is:

$$Y = \frac{Cv_1 vB_1}{\lambda_1}(e^{\lambda_1 t} - 1) + \frac{Cv_2 vB_2}{\lambda_2}(e^{\lambda_2 t} - 1).$$

The data needed are in the following table.

Data Needed for Step Response Using λ_1 = -1.0025 and λ_2 = -1.9975			
$vB = \begin{bmatrix} -1.9 \\ 3.9 \end{bmatrix}$	$Cv = \begin{bmatrix} -0.975 \\ -0.475 \end{bmatrix}^T$	$\dfrac{Cv_1 vB_1}{\lambda_1} = -1.848$	$\dfrac{Cv_2 vB_2}{\lambda_2} = 0.927$

Plugging data, we get:

$$Y = -1.848(e^{-1.0025t} - 1) + 0.927(e^{-1.9975t} - 1).$$

The following figure compares Y with the results of *Runge-Kutta* integration. There is an error.

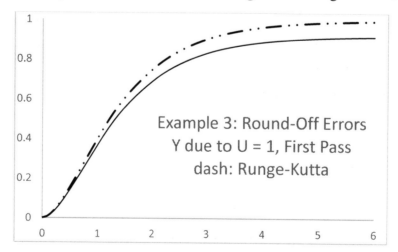

Example 3: Round-Off Errors
Y due to U = 1, First Pass
dash: Runge-Kutta

Let's now compute a new set of eigenvectors using the diagonal elements of the L matrix as *new* eigenvalues.

The eigenvector for $\lambda_1 = -1.0025$ is computed as shown in the following table. The strategy is the same.

Form the Base Equations	Pick $x_2 = 1$	The Eigenvector From (E2)
(E1) $1.0025x_1 + x_2 = 0$	Compute $x_1 = -0.9975$	$\bar{x} = \begin{bmatrix} -0.9988 \\ 1 \end{bmatrix}$
(E2) $-2x_1 - 1.9975x_2 = 0$	Compute $x_1 = -0.9988$	

The eigenvector for $\lambda_1 = -1.9975$ is computed as shown in the following table.

Form The Base Equations	Pick $x_2 = 1$	The Eigenvector From (E2)
(E1) $1.9975x_1 + x_2 = 0$	Compute $x_1 = -0.5006$	$\bar{x} = \begin{bmatrix} -0.5013 \\ 1 \end{bmatrix}$
(E2) $-2x_1 - 1.0025x_2 = 0$	Compute $x_1 = -0.5013$	

The combined matrices are shown in the following table.

Combined Matrices
$V = \begin{bmatrix} -0.9988 & -0.5013 \\ 1 & 1 \end{bmatrix}$ $V^{-1} = \begin{bmatrix} -2.01 & -1.0075 \\ 2.01 & 2.0075 \end{bmatrix}$ $L = V^{-1}AV = \begin{bmatrix} -1 & 0.0025 \\ -0.0025 & -2 \end{bmatrix}$

For the L matrix, the off-diagonal elements are smaller than the *first try*, but are they small enough? We could *try again*, or we could compute a step response.

The data to do this are shown in the following table.

Data Needed for Step Response Using $\lambda_1 = -1$ and $\lambda_2 = -2$ From the Diagonal of L
$vB = \begin{bmatrix} -2.015 \\ 4.015 \end{bmatrix}$ $Cv = \begin{bmatrix} -0.999 \\ -0.501 \end{bmatrix}^T$ $\dfrac{Cv_1 vB_1}{\lambda_1} = -2.013$ $\dfrac{Cv_2 vB_2}{\lambda_2} = 1.006$

Plugging these numbers, we can form the following equation:

$$Y = -2.013(e^{-t} - 1) + 1.006(e^{-2t} - 1).$$

The following figure compares this Y with the results of *Runge-Kutta* integration.

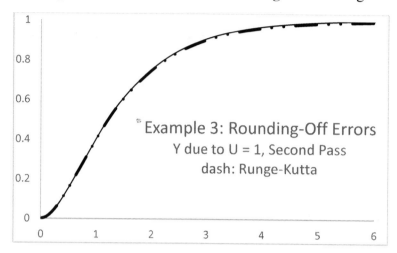

Example 3: Rounding-Off Errors
Y due to U = 1, Second Pass
dash: Runge-Kutta

We have shown that we can repeatedly compute eigenvectors until the off-diagonal elements of matrix L are *sufficiently small*. When we write a program to compute eigenvectors, this *feedback* will become part of the strategy.

Examples

Example 4: Eigenvectors Used as Mode Shapes

The field of structural dynamics uses eigenvectors as mode shapes. We will define and demonstrate this using the system equations from **example 2**.

$$\dot{Z}_1 = Z_2$$
$$\dot{Z}_2 = 2*(Z_3 - Z_1) - 0.5Z_2 \text{ and } Y = Z_1$$
$$\dot{Z}_3 = 10*(U - Z_3)$$

If $U = 1$, in steady state, $\dot{Z}_1 = \dot{Z}_2 = \dot{Z}_3 = 0$. This requires that $Z_1 = Z_3 = U = 1$. The problem in this example is to determine the response when U is removed.

From **example 2**, the response is composed of two modes:

- A decaying exponential mode with an eigenvalue of -10.

- An exponentially decaying sinusoidal mode with eigenvalues of $-0.25 + j1.39$ and $-0.25 - j1.39$:

$$Y = Z_1 = A_1 e^{-10t} + e^{-0.25t}\left[A_2 \cos(1.39t) + B_2 \sin(1.39t)\right].$$

The problem is to determine the values of A_1, A_2, and B_2.

At $t = 0$, $\qquad\qquad\qquad Z_1 = 1 = A_1 + A_2.$ $\qquad\qquad\qquad$ Equation A

We need more equations. The output rate is:

$$\dot{Z}_1 = -10*A_1 e^{-10t} + e^{-0.25t}\left[-1.39 A_2 \sin(1.39t) + 1.39 B_2 \cos(1.39t)\right]$$
$$- 0.25 e^{-0.25t}\left[A_2 \cos(1.39t) + B_2 \sin(1.39t)\right]$$

Since at $t = 0$, $\dot{Z}_1 = 0$, we have another equation: $\quad 0 = -10 A_1 + 1.39 B_2 - 0.25 A_2$ $\qquad$ Equation B

From **example 2**, we have $\begin{bmatrix} Z_1 \\ Z_2 \\ Z_3 \end{bmatrix} = V \begin{bmatrix} q_1 \\ q_2 \\ q_3 \end{bmatrix}$, where V is the matrix of eigenvectors.

$$\begin{bmatrix} Z_1 \\ Z_2 \\ \hline Z_3 \end{bmatrix} = \begin{bmatrix} 0.0206 & -0.125-j0.696 & -0.125+j0.696 \\ -0.206 & 1 & 1 \\ \hline 1 & 0 & 0 \end{bmatrix} \begin{bmatrix} q_1 \\ q_2 \\ q_3 \end{bmatrix}.$$

The partitioning emphasizes that Z_1 and Z_2 are combined.

We see that $Z_3 = q_1$, which is the coordinate for $\lambda = -10$. Therefore, $Z_3 = A_3 e^{-10t}$.

At $t = 0$: $\qquad\qquad\qquad Z_3 = 1 = A_3.$ $\qquad\qquad\qquad$ Equation C

Let's look at the eigenvector for λ = -10. It's: $\begin{bmatrix} 0.0206 \\ -0.206 \\ \hline 1 \end{bmatrix}$. The ratio of Z_3 to Z_1 is $\dfrac{1}{0.0206} = 48.54$

This means that $A_3 = 48.54 A_1$. Equation C becomes: $Z_3 = 1 = 48.54 A_1$. Equations A, B, and C yield:

$$A_1 = 0.0206, \quad A_2 = 0.9794, \quad \text{and} \quad B_2 = 0.324.$$

Therefore: $\quad Y = 0.0206 e^{-10t} + e^{-0.25t}\left[0.9794\cos(1.39t) + 0.324\sin(1.39t)\right].$

The following figure is a plot of Y. Comparison with Runge-Kutta integration is perfect.

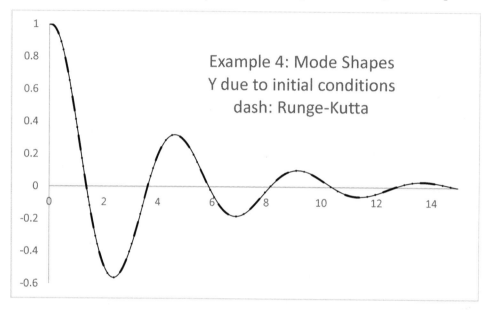

Example 4: Mode Shapes
Y due to initial conditions
dash: Runge-Kutta

We have used the eigenvector for λ = -10 as a mode shape. In case 4 of chapter 15, eigenvectors are used to show the shape that a structure will take as it vibrates at each of its internal frequencies or eigenvalues.

Note: The coefficients in the equation for Y *match* those in the step response of example 2.

Chapter 2: A Program to Compute Eigenvalues of a General Real Matrix

The eigenvalues of the matrix A are computed from the *characteristic* equation: $|A - \lambda I| = 0$. This is a polynomial equation. In general, factoring this polynomial is not the way to compute eigenvalues. Francis developed the following iterative method.[2]

For the moment, assume that the eigenvalues λ are real numbers and that A is in the upper-triangular form:

$$|A - \lambda I| = \begin{vmatrix} (a_{11}-\lambda) & a_{12} & a_{13} & \bullet & a_{1n} \\ 0 & (a_{22}-\lambda) & a_{23} & \bullet & a_{2n} \\ 0 & 0 & (a_{33}-\lambda) & \bullet & a_{3n} \\ \bullet & \bullet & \bullet & \bullet & \bullet \\ 0 & 0 & 0 & \bullet & (a_{nn}-\lambda) \end{vmatrix}.$$

This determinant can be evaluated by expanding in minors. By inspection:

$$|A - \lambda I| = (a_{11}-\lambda)(a_{22}-\lambda) \bullet \bullet (a_{nn}-\lambda).$$

The λ's of an upper-triangular matrix are the elements on its diagonal. Rutishauser discovered how to iteratively transform a general matrix into upper-triangular form while preserving its eigenvalues.[3] This process is known as the LR algorithm.

Starting with a matrix A_o, the first step is to factor it into a lower-triangular matrix L_o and an upper-triangular matrix R_o:

$$A_o = L_o * R_o.$$

I discuss later how to compute L_o and R_o. Because they have the same λ's, A_o and $(L_o * R_o)$ are similar. From the definition of similarity, this means that for *any* matrix T:

$$T * A_o = (L_o * R_o) * T.$$

Using L_o as the T matrix:

$$L_o * A_o = (L_o * R_o) * L_o.$$

Then:

$$A_o = L_o^{-1} * (L_o * R_o) * L_o = R_o * L_o.$$

This new matrix $(R_o * L_o)$ comes from multiplying the factors of A_o in reverse. It has the same eigenvalues as the original A_o matrix. The LR algorithm now uses this new matrix as the starting point for the next iteration. This is shown in the following figure.

---

[2] J. Francis, *The QR Transformation, Parts I and II* (Computer Journal, 1961).
[3] H. Rutishauser, *Solution of Eigenvalue Problems with the LR Transformation* (National Bureau of Standards Applied Math Series, 1958).

Eigenvalues

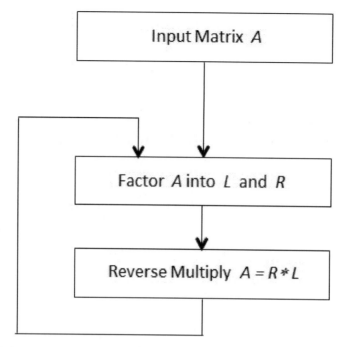

Rutishauser proved that if you continue this process and there are no numerical difficulties, the following will happen. The matrix L will approach the identity matrix. When this happens, A equals R and is therefore upper triangular. Its eigenvalues are the same as the original A matrix and are on its diagonal.

Appendix D.1 contains a program written in the VBA language that implements the LR algorithm. The reader will be able to understand and use it after reading chapter 4 through 13. For now, I illustrate its use in the following example.

Example Demo 1: Use the LR program to verify that the λ's of the following matrix are $\lambda = 4, 6,$ and 12:

$$A = \begin{bmatrix} 6 & 2 & -2 \\ 2 & 6 & -2 \\ -2 & -2 & 10 \end{bmatrix}.$$

The following table shows the results.

Eigenvalues

Iteration (k)	The A Matrices at Start of Each Iteration (The eigenvalues are $\lambda_1 = 4$, $\lambda_2 = 6$, and $\lambda_3 = 12$ for each of these matrices.)			The L Matrices (lower triangular)			The R Matrices (upper triangular)		
0	6	2	-2	1	0	0	6	2	-2
	2	6	-2	0.33	1	0	0	5.33	-1.33
	-2	-2	10	-0.33	-0.25	1	0	0	9
1	7.33	2.5	-2	1	0	0	7.33	2.5	-2
	2.22	5.67	-1.33	0.3	1	0	0	4.91	-0.73
	-3.00	-2.25	9	-0.41	-0.25	1	0	0	8
2	8.91	3	-2	1	0	0	8.91	3	-2
	1.79	5.09	-0.73	0.2	1	0	0	4.49	-0.33
	-3.27	-2	8	-0.37	-0.2	1	0	0	7.2
3	10.24	3.4	-2	1	0	0	10.24	3.4	-2
	1.02	4.56	-0.33	0.1	1	0	0	4.22	-0.13
	-2.64	-1.44	7.2	-0.26	-0.13	1	0	0	6.67
4	11.10	3.67	-2	1	0	0	11.1	3.67	-2
	0.45	4.23	-0.13	0.04	1	0	0	4.08	-0.05
	-1.72	-0.89	6.67	-0.16	-0.08	1	0	0	6.35
5	11.56	3.82	-2	1	0	0	11.56	3.82	-2
	0.17	4.09	-0.05	0.02	1	0	0	4.03	-0.02
	-0.99	-0.5	6.35	-0.09	-0.04	1	0	0	6.18
6	11.79	3.91	-2	1	0	0	11.79	3.91	-2
	0.06	4.03	-0.02	0.01	1	0	0	4.01	-0.01
	-0.53	-0.26	6.18	-0.04	-0.02	1	0	0	6.09
7	11.9	3.95	-2	1	0	0	11.9	3.95	-2
	0.02	4.01	-0.01	0	1	0	0	4	0
	-0.27	-0.14	6.09	-0.02	-0.01	1	0	0	6.05
8	11.95	3.98	-2	1	0	0	11.95	3.98	-2
	0.01	4	0	0	1	0	0	4	0
	-0.14	-0.07	6.05	-0.01	-0.01	1	0	0	6.02
9	11.98	3.99	-2	1	0	0	11.98	3.99	-2
	0	4	0	0	1	0	0	4	0
	-0.07	-0.03	6.02	-0.01	0	1	0	0	6.01
10	11.99	3.99	-2	1	0	0	11.99	3.99	-2
	0	4	0	0	1	0	0	4	0
	-0.04	-0.02	6.01	0	0	1	0	0	6.01
11	11.99	4	-2	1	0	0	11.99	4	-2
	0	4	0	0	1	0	0	4	0
	-0.02	-0.01	6.01	0	0	1	0	0	6
12	12	4	-2	1	0	0	12	4	-2
	0	4	0	0	1	0	0	4	0
	-0.01	0	6	0	0	1	0	0	6
13	12	4	-2						
	0	4	0						
	0	0	6						

Table title: LR Program Results for Example Demo 1 (*Note:* $A_k = L_k * R_k$ and $A_{k+1} = R_k * L_k$)

By iteration 13, A has been triangularized, and the λ's are seen on its diagonal. Note that the zeros in most of the matrices are actually small numbers less than 1e-4.

Example Demo 2: Use the LR program to verify that the λ's of the following matrix are λ = 1, 2, and 5:

$$A = \begin{bmatrix} 1 & -1 & 1 \\ 4 & 6 & -1 \\ 4 & 4 & 1 \end{bmatrix}.$$

The following table shows the results.

Iteration (k)	The A Matrices at Start of Each Iteration (The eigenvalues are $\lambda_1 = 1$, $\lambda_2 = 2$, and $\lambda_3 = 5$ for each of these matrices.)			The L Matrices (lower triangular)			The R Matrices (upper triangular)		
0	1	-1	1	1	0	0	1	-1	1
	4	6	-1	4	1	0	0	10	-5
	4	4	1	4	0.80	1	0	0	1
1	1	-0.20	1	1	0	0	1	-0.20	1
	20	6	-5	20	1	0	0	10	-25
	4	0.80	1	4	0.16	1	0	0	1
2	1	-0.04	1	1	0	0	1	-0.04	1
	100	6	-25	100	1	0	0	10	-125
	4	0.16	1	4	0.03	1	0	0	1
3	1	-0.01	1	1	0	0	1	-0.01	1
	500	6	-125	500	1	0	0	10	-625
	4	0.03	1	4	0.01	1	0	0	1
4	1	0	1	1	0	0	1	0	1
	2500	6	-625	2500	1	0	0	10	-3125
	4	0.01	1	4	0	1	0	0	1
5	1	0	1						
	12500	6	-3125						
	4	0	1						

LR Program Results for Example Demo 2 (*Note:* $A_k = L_k * R_k$ and $A_{k+1} = R_k * L_k$)

The program is diverging. Division by small numbers has been encountered. The LR algorithm implicitly involves matrix inversion.

The method by Francis, previously referenced, avoids matrix inversion by replacing L with an orthogonal matrix. The inverse of an orthogonal matrix is its transpose. Francis called this the Q matrix: $A = Q * R$, where R is still upper triangular. This kind of factorization is discussed in a work by Hager.[4] The Francis algorithm is otherwise exactly like the LR, and it is called the QR. A program that implements the QR algorithm is shown in appendix D.2.

---

[4] W. Hager, *Applied Numerical Linear Algebra* (Upper Saddle River, NJ: Prentice Hall, 1988).

Let's use this program on the same matrix where the LR algorithm failed.

Example Demo 2 using the QR program: The following table shows the results.

QR Program Results for Example Demo 2 (*Note:* $A_k = Q_k * R_k$ and $A_{k+1} = R_k * Q_k$)									
Iteration (k)	The A Matrices at Start of Each Iteration (The eigenvalues are $\lambda_1 = 1$, $\lambda_2 = 2$, and $\lambda_3 = 5$ for each of these matrices.)			The Q Matrices (orthogonal)			The R Matrices (upper triangular)		
0	1	-1	1	-0.17	0.83	-0.53	-5.74	-6.79	-0.17
	4	6	-1	-0.70	-0.48	-0.53	0	-2.63	1.59
	4	4	1	-0.70	0.28	0.66	0	0	0.66
1	5.85	-1.53	6.53	-0.99	0.12	0.08	-5.91	1.32	-6.72
	0.72	1.71	2.45	-0.12	-0.99	-0.03	0	-1.89	-1.67
	-0.46	0.18	0.44	0.08	-0.04	1	0	0	0.90
2	5.16	-1.73	-7.23	-1	0.02	-0.01	-5.16	1.69	7.24
	0.10	1.94	-1.60	-0.02	-1	0.01	0	-1.98	1.47
	0.07	-0.04	0.89	-0.01	0.01	1	0	0	0.98
3	5.03	-1.73	7.33	-1	0	0	-5.03	1.72	-7.33
	0.02	1.99	1.45	0	-1	0	0	-2	-1.42
	-0.01	0.01	0.98	0	0	1	0	0	1
4	5.01	-1.73	-7.34	-1	0	0	-5.01	1.73	7.34
	0	2	-1.42	0	-1	0	0	-2	1.42
	0	0	1	0	0	1	0	0	1
5	5	-1.73	7.35						
	0	2	1.42						
	0	0	1						

The QR program works.

Example Demo 3: Apply the QR program to the problem in **Example Demo 1** that used the LR. The following table shows the results.

Eigenvalues

Iteration (k)	The A Matrices at Start of Each Iteration (The eigenvalues are $\lambda_1 = 4$, $\lambda_2 = 6$, and $\lambda_3 = 12$ for each of these matrices.)			The Q Matrices (orthogonal)			The R Matrices (upper triangular)		
0	6	2	-2	-0.9	0.36	0.24	-6.63	-4.22	5.43
	2	6	-2	-0.3	-0.92	0.24	0	-5.12	2.56
	-2	-2	10	0.3	0.14	0.94	0	0	8.49
1	8.91	2.31	2.56	-0.93	0.26	-0.25	-9.55	-3.71	-4.82
	2.31	5.09	1.21	-0.24	-0.96	-0.12	0	-4.35	-0.87
	2.56	2.21	8	-0.27	-0.05	0.96	0	0	6.93
2	11.10	1.29	-1.86	-0.98	0.12	0.16	-11.33	-1.8	2.95
	1.29	4.23	-0.34	-0.11	-0.99	0.03	0	-4.06	0.19
	-1.86	-0.34	6.67	0.16	0.01	0.99	0	0	6.27
3	11.79	0.49	1.03	-1	0.04	-0.09	-11.84	-0.66	-1.56
	0.49	4.03	0.06	-0.04	-1	-0.01	0	-4.01	-0.03
	1.03	0.06	6.18	-0.09	0	1	0	0	6.07
4	11.95	0.17	-0.53	-1	0.01	0.04	-11.96	-0.23	0.79
	0.17	4	-0.01	-0.01	-1	0	0	-4	0.01
	-0.53	-0.01	6.05	0.04	0	1	0	0	6.02
5	11.99	0.06	0.26	-1	0	-0.02	-11.99	-0.08	-0.4
	0.06	4	0	0	-1	0	0	-4	0
	0.26	0	6.01	-0.02	0	1	0	0	6
6	12	0.02	-0.13	-1	0	0.01	-12	-0.03	0.2
	0.02	4	0	0	-1	0	0	-4	0
	-0.13	0	6	0.01	0	1	0	0	6
7	12	0.01	0.07	-1	0	-0.01	-12	-0.01	-0.1
	0.01	4	0	0	-1	0	0	-4	0
	0.07	0	6	-0.01	0	1	0	0	6
8	12	0	-0.03	-1	0	0	-12	0	0.05
	0	4	0	0	-1	0	0	-4	0
	-0.03	0	6	0	0	1	0	0	6
9	12	0	0.02	-1	0	0	-12	0	-0.02
	0	4	0	0	-1	0	0	-4	0
	0.02	0	6	0	0	1	0	0	6
10	12	0	-0.01	-1	0	0	-12	0	0.01
	0	4	0	0	-1	0	0	-4	0
	-0.01	0	6	0	0	1	0	0	6
11	12	0	0						
	0	4	0						
	0	0	6						

QR Program Results for Example Demo 3 (*Note:* $A_k = Q_k * R_k$ and $A_{k+1} = R_k * Q_k$)

By iteration 11, A has been successfully triangularized. Note that the off-diagonal terms for the final A are different from those from the LR program.

Complex Eigenvalues

For real matrices, complex eigenvalues come in conjugate pairs. It takes four elements in a matrix of real numbers to yield one complex pair. For example:

$$A = \begin{bmatrix} a & b \\ c & d \end{bmatrix}.$$

The eigenvalues of A are the roots of:

$$\begin{vmatrix} (a-\lambda) & b \\ c & (d-\lambda) \end{vmatrix} = 0.$$

Assume that a matrix has four real eigenvalues and two complex pairs. The LR and QR algorithms must iterate until the form of the matrix resembles the following.

$$\begin{bmatrix} x & x & x & x & x & x & x & x \\ x & x & x & x & x & x & x & x \\ & x & x & x & x & x & x & x \\ & & x & x & x & x & x & x \\ & & & & x & x & x & x \\ & & & & & x & x & x \\ & & & & & x & x & x \\ & & & & & & & x \end{bmatrix}.$$

As can be seen, the complex eigenvalues show up as *bulges* on the diagonal. Detecting complex roots is a big part of the LR and QR convergence criteria. The following example demonstrates complex eigenvalues.

Example Demo 4: Apply the QR program to the following matrix:

$$A = \begin{bmatrix} 0 & 1 & 0 \\ -6 & -2 & 0.0027 \\ -243 & -32 & 0.028 \end{bmatrix}.$$

Its eigenvalues are: $\lambda_1 = -0.08$, $\lambda_2 = -0.94 + j2.12$, $\lambda_3 = -0.94 - j2.12$. The following table shows the results.

Eigenvalues

Iteration (k)	The A Matrices at Start of Each Iteration (The eigenvalues are $\lambda_1 = -.08$, $\lambda_2 = -0.94+j2.12$, $\lambda_3 = -0.94-j2.12$ for each of these matrices.)			The Q Matrices (orthogonal)			The R Matrices (upper triangular)		
0	0 -6 -243	1 -2 -32	0 0.0027 0.028	-0.01 1 0	-0.78 -0.01 0.63	0.63 0.01 0.78	-243.01 0 0	1.19 -0.04 0	-32.01 -0.11 1.05
1	1.19 -0.04 0	168.43 -2.02 0.66	-178.04 -0.08 -1.14	-1 0.02 0	-0.02 -0.96 -0.28	-0.01 -0.28 0.96	-2.4 0 0	-168.4 -2.38 0	178.01 3.2 1
2	-1.9 -0.04 0	112.34 0.18 -0.28	217.8 3.74 -0.25	-1 -0.01 0	0.01 -1 -0.09	0 -0.09 1	3.86 0 0	-112.3 3.06 0	-217.8 -1.07 -2.32
3	-0.63 -0.03 0	131.6 -0.99 0.21	-206.77 -1.34 -0.35	-1 -0.01 0	0.01 -1 0.04	0 0.04 1	2.7 0 0	-131.6 4.75 0	206.77 -1.42 -2.36
4	1.06 -0.06 0	140.56 -2.74 -0.1	200.77 -1.21 -0.29	-1 0.05 0	-0.05 -1 0.03	0 0.03 1	-1.3 0 0	-140.5 -4.11 0	-200.6 -8.23 0.16
5	-5.55 -0.19 0	135.31 3.66 0	-204.09 -8.34 -0.08						

QR Program Results for Example Demo 4 (*Note:* $A_k = Q_k * R_k$ and $A_{k+1} = R_k * Q_k$)

At iteration 5, the last row shows $\lambda_1 = -0.08$. The *bulge* shows the other two λ's. They are the roots of the following equation:

$$\begin{vmatrix} (-5.55-\lambda) & 135.31 \\ -0.19 & (3.66-\lambda) \end{vmatrix} = 0.$$

The solutions of this equation are: $\lambda_2 = -0.94 + j2.12$, $\lambda_3 = -0.94 - j2.12$.

The QR Algorithm Expanded to Include *Matrix Deflation*

Suppose the following A matrix is detected during the iterations:

$$\begin{bmatrix} E & | & F \\ \hline 0 \bullet\bullet\bullet 0 & | & a \end{bmatrix} \qquad Condition\ 1.$$

Here, the last row reveals that an eigenvalue has been found. When this matrix is expanded by minors, the last row and column can be dropped. QR iterations can continue with matrix A = E. This is called *matrix deflation*.

Now suppose the following A matrix is detected during the iterations:

$$\begin{bmatrix} G & | & H \\ \hline 0 \bullet\bullet\bullet 0 & | & a & b \\ 0 \bullet\bullet\bullet 0 & | & c & d \end{bmatrix} \qquad Condition\ 2.$$

Here the last two rows reveal a pair of eigenvalues. After computing these λ's, the last two rows and columns can be dropped, and the QR can continue with A = G.

The QR Algorithm Expanded to Include *Shifting*

Bronson discusses a theorem that says that if A has an eigenvalue λ_1, then one of the eigenvalues of the matrix $[A - \sigma_1 I]$ will be $(\lambda_1 - \sigma_1)$.[5] This is called *shifting*. If $\sigma_1 = \lambda_1$, then the matrix $[A - \lambda_1 I]$ will have an eigenvalue of zero. When this matrix is factored, the last row of R is forced to have all zeros. The last row of the reverse multiplication will also be all zeros. To regain similarity, the *shift* of λ_1 must be added back in. This is done via the statement $A = [A + \lambda_1 I]$. Hence the eigenvalue λ_1 will be revealed in the last row of A. The following example shows this.

Example Demo 5: This example will use the matrix in **Example Demo 3**:

$$\begin{bmatrix} 6 & 2 & -2 \\ 2 & 6 & -2 \\ -2 & -2 & 10 \end{bmatrix}.$$

- Its eigenvalues are $\lambda = 4, 6,$ and 12. Set $\text{shift} = 12 * I$.

- Perform the shift, $A = [A - 12 * I]$:

$$\begin{bmatrix} -6 & 2 & -2 \\ 2 & -6 & -2 \\ -2 & -2 & -2 \end{bmatrix}.$$

- Factor the shifted A into Q and R:

$$Q = \begin{bmatrix} -0.9 & -0.12 & 8 \\ 0.3 & -0.86 & -6 \\ -0.3 & -0.49 & 0 \end{bmatrix} \quad R = \begin{bmatrix} 6.63 & -3.02 & 1.8 \\ 0 & 5.91 & 2.95 \\ 0 & 0 & 0 \end{bmatrix}.$$

- Reverse multiply $A = R * Q$:

$$\begin{bmatrix} -1.45 & 0.89 & 71.16 \\ 0.89 & -6.55 & -35.45 \\ 0 & 0 & 0 \end{bmatrix}.$$

- To restore similarity, shift back $A = [A + 12 * I]$:

$$\begin{bmatrix} 4.55 & 0.89 & 71.16 \\ 0.89 & 5.45 & -35.45 \\ 0 & 0 & 12 \end{bmatrix}.$$

- From the last row, $\lambda_1 = 12$.

---

[5] R. Bronson, *Linear Algebra* (Cambridge, MA: Academic Press, 1995).

- From the bulge on the diagonal, the remaining λ's are the roots of:

$$\begin{vmatrix} (4.55-\lambda) & 0.89 \\ 0.89 & (5.45-\lambda) \end{vmatrix} = 0 \quad \text{which are } \lambda_2 = 4 \text{ and } \lambda_3 = 6.$$

With *matrix deflation* and *shifting*, the QR takes one iteration. From **Example Demo 3**, it takes 11 iterations without them.

Of course, exact shifting is not feasible because we don't know beforehand where the λ's are. But experience has led to the development of several successful *guessing methods for shifting*.

Francis proposed the following strategy: eigenvalues are computed from the submatrix in the lower-right-hand corner of the A matrix.

$$\begin{bmatrix} a_{n-1,n-1} & a_{n-1,n} \\ a_{n,n-1} & a_{n,n} \end{bmatrix} \qquad \textit{Guess 1}.$$

Pick the eigenvalue of this submatrix that is closest to the element $a_{n,n}$. Use this eigenvalue as the *shift* for each QR iteration. The *shift* changes on each iteration, and it will become closer to a real eigenvalue of A. If the next eigenvalue is complex, remember that *shifting* is only an aid.

There is another benefit from *shifting*. The QR does not diverge like the LR, but it can get into a limit-cycle. Sometimes after several iterations, the QR matrices begin to repeat. When this happens, there will be no convergence. Francis discovered that this limit-cycle could be interrupted by changing the method of *shifting* during the iteration process. His method is complicated.

I use a simpler method that is remarkably effective. If the first guess doesn't yield convergence after an arbitrary 100 iterations, the submatrix used to compute the *shift* is changed:

$$\begin{bmatrix} a_{n-1,n-2} & a_{n-1,n} \\ a_{n,n-2} & a_{n,n} \end{bmatrix} \qquad \textit{Guess 2}.$$

The QR Algorithm Expanded to Include the Hessenberg Matrix

There is one final step that has become a part of most QR strategies. A Hessenberg matrix has all zeros below the subdiagonal:

$$\begin{bmatrix} x & x & x & x & x & x \\ x & x & x & x & x & x \\ 0 & x & x & x & x & x \\ 0 & 0 & x & x & x & x \\ 0 & 0 & 0 & x & x & x \\ 0 & 0 & 0 & 0 & x & x \end{bmatrix}.$$

Being nearly triangular, it is common to transform the initial A to Hessenberg form before beginning QR iterations. Scheid discusses this transformation, which maintains similarity.[6]

---

[6] F. Scheid, *Numerical Analysis* (McGraw-Hill, 1988).

The following is a flowchart of the program with all the expansions.

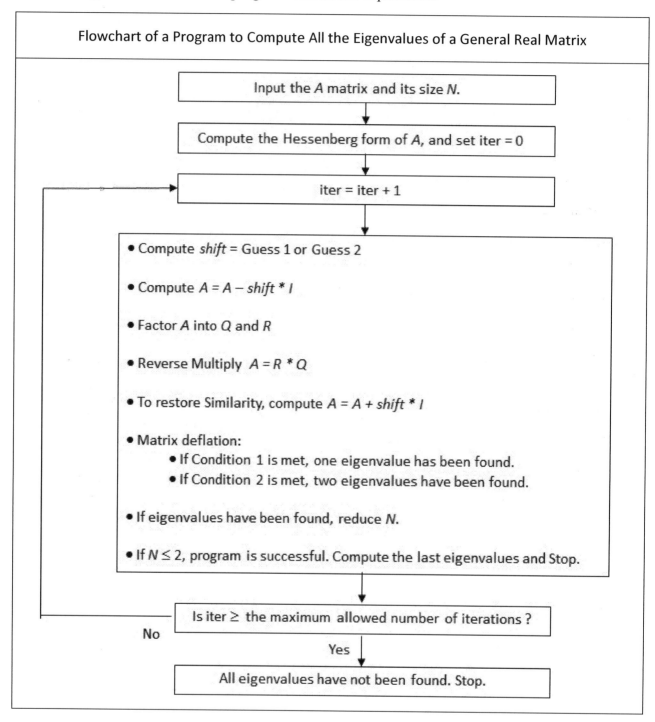

The program that implements these steps will be shown in chapter 14, after the chapters on the VBA programming language.

Chapter 3: A Program to Compute Eigenvectors of a General Real Matrix

An eigenvector $\bar{x}$ is computed from the following equation:

$$[A - \lambda I][\bar{x}] = \begin{bmatrix} (a_{11}-\lambda) & a_{12} & \cdots & a_{1n} \\ a_{21} & (a_{22}-\lambda) & \cdots & a_{2n} \\ \vdots & \vdots & \cdots & \vdots \\ a_{n1} & a_{n2} & \cdots & (a_{nn}-\lambda) \end{bmatrix} \begin{bmatrix} x_1 \\ x_2 \\ \vdots \\ x_n \end{bmatrix} = 0.$$ The Base Equation

- This is a system of n equations, n unknowns, and an infinite number of solutions.

- We are free to pick a value for one of the unknowns and *solve for the others.*

- If there are no numerical difficulties and if all a_{ij} are nonzero, we can solve for the others using *any* (n – 1) subset of the equations.

- But mostly, some a_{ij} will be zero. This means we must search for the subset to use.

- Besides searching for the subset, for which variable are we free to pick the value?

- This is a perfect problem for an exhaustive search. Find the subset and the variable by trying all the possibilities.

Let's demonstrate this.

Given: $A = \begin{bmatrix} -6 & 6 & 0 \\ 0 & -4 & 4 \\ 0 & 0 & -2 \end{bmatrix}$. The eigenvalues are: $\lambda = -6, -4,$ and -2.

The following are the base equations:
E 1) $(-6-\lambda)x_1 + 6x_2 + 0x_3 = 0$
E 2) $0x_1 + (-4-\lambda)x_2 + 4x_3 = 0$.
E 3) $0x_1 + 0x_2 + (-2-\lambda)x_3 = 0$

- For the eigenvalue $\lambda = -6$, the base equations are:
E 1) $0x_1 + 6x_2 + 0x_3 = 0$
E 2) $0x_1 + 2x_2 + 4x_3 = 0$.
E 3) $0x_1 + 0x_2 + 4x_3 = 0$

If we pick a value for any x_i, there are three possible subsets. With three variables, there are nine possible searches. These are summarized in the following table, starting with $x_3 = 1$.

Base Equations	Subset 1	Subset 2	Subset 3
When $x_3 = 1$ E 1) $0x_1 + 6x_2 = 0$ E 2) $0x_1 + 2x_2 = -4$ E 3) $0x_1 + 0x_2 = -4$	1 E 2) $0x_1 + 2x_2 = -4$ E 3) $0x_1 + 0x_2 = -4$ No Solution	2 E 1) $0x_1 + 6x_2 = 0$ E 3) $0x_1 + 0x_2 = -4$ No Solution	3 E 1) $0x_1 + 6x_2 = 0$ E 2) $0x_1 + 2x_2 = -4$ No Solution
When $x_2 = 1$ E 1) $0x_1 + 0x_3 = -6$ E 2) $0x_1 + 4x_3 = -2$ E 3) $0x_1 + 4x_3 = 0$	4 E 2) $0x_1 + 4x_3 = -2$ E 3) $0x_1 + 4x_3 = 0$ No Solution	5 E 1) $0x_1 + 0x_3 = -6$ E 3) $0x_1 + 4x_3 = 0$ No Solution	6 E 1) $0x_1 + 0x_3 = -6$ E 2) $0x_1 + 4x_3 = -2$ No Solution
When $x_1 = 1$ E 1) $6x_2 + 0x_3 = 0$ E 2) $2x_2 + 4x_3 = 0$ E 3) $0x_2 + 4x_3 = 0$	7 E 2) $2x_2 + 4x_3 = 0$ E 3) $0x_2 + 4x_3 = 0$ $x_2 = 0$ and $x_3 = 0$	8 Not Needed	9 Not Needed

On the seventh try, the eigenvector for $\lambda = -6$ is $\bar{x} = \begin{bmatrix} 1 & 0 & 0 \end{bmatrix}^T$.

- For the eigenvalue $\lambda = -4$, the base equations are:
  E 1) $-2x_1 + 6x_2 + 0x_3 = 0$
  E 2) $0x_1 + 0x_2 + 4x_3 = 0$
  E 3) $0x_1 + 0x_2 + 2x_3 = 0$

The following table shows the search.

Base Equations	Subset 1	Subset 2	Subset 3
When $x_3 = 1$ E 1) $-2x_1 + 6x_2 = 0$ E 2) $0x_1 + 0x_2 = -4$ E 3) $0x_1 + 0x_2 = -2$	1 E 2) $0x_1 + 0x_2 = -4$ E 3) $0x_1 + 0x_2 = -2$ No solution	2 E 1) $-2x_1 + 6x_2 = 0$ E 3) $0x_1 + 0x_2 = -2$ No solution	3 E 1) $-2x_1 + 6x_2 = 0$ E 2) $0x_1 + 0x_2 = -4$ No solution
When $x_2 = 1$ E 1) $-2x_1 + 0x_3 = -6$ E 2) $0x_1 + 4x_3 = 0$ E 3) $0x_1 + 2x_3 = 0$	4 E 2) $0x_1 + 4x_3 = 0$ E 3) $0x_1 + 2x_3 = 0$ No solution	5 E 1) $-2x_1 + 0x_3 = -6$ E 3) $0x_1 + 2x_3 = 0$ $x_1 = 3$ and $x_3 = 0$	6 Not Needed

On the fifth try, the eigenvector for $\lambda = -4$ is: $\bar{x} = \begin{bmatrix} 3 & 1 & 0 \end{bmatrix}^T$. We are free to normalize this to

$$\bar{x} = \begin{bmatrix} 1 & 1/3 & 0 \end{bmatrix}^T.$$

Eigenvectors

- For the eigenvalue $\lambda = -2$, the base equations are:
$$\begin{aligned} E\ 1)\ -4x_1 + 6x_2 + 0x_3 &= 0 \\ E\ 2)\ 0x_1 - 2x_2 + 4x_3 &= 0 \\ E\ 3)\ 0x_1 + 0x_2 + 0x_3 &= 0 \end{aligned}$$

The following table shows the search.

Base Equations	Subset 1	Subset 2	Subset 3
When $x_3 = 1$ E 1) $-4x_1 + 6x_2 = 0$ E 2) $0x_1 - 2x_2 = -4$ E 3) $0x_1 + 0x_2 = 0$	1 E 2) $0x_1 - 2x_2 = -4$ E 3) $0x_1 + 0x_2 = 0$ No solution	2 E 1) $-4x_1 + 6x_2 = 0$ E 3) $0x_1 + 0x_2 = 0$ No solution	3 E 1) $-4x_1 + 6x_2 = 0$ E 2) $0x_1 - 2x_2 = -4$ $x_1 = 3$ and $x_2 = 2$

The eigenvector for $\lambda = -2$ is: $\overline{x} = \begin{bmatrix} 3 & 2 & 1 \end{bmatrix}^T$. We can make this $\overline{x} = \begin{bmatrix} 1 & 2/3 & 1/3 \end{bmatrix}^T$.

To computerize this search, three flags will be needed:

- **op** will identify which variable is picked.
- **pass** will identify which subset is being tried.
- A third flag, **irow**, will be discussed shortly.

The following table summarizes this.

Summary of the Exhaustive Search When the Matrix Size Is Three and *irow = 1*			
Base Equations	Subset 1	Subset 2	Subset 3
When $x_3 = 1$ (op = 1) Equation 1 Equation 2 Equation 3	1. pass = 1 Equation 2 Equation 3	2. pass = 2 Equation 1 Equation 3	3. pass = 3 Equation 1 Equation 2
When $x_2 = 1$ (op = 2) Equation 1 Equation 2 Equation 3	4. pass = 1 Equation 2 Equation 3	5. pass = 2 Equation 1 Equation 3	6. pass = 3 Equation 1 Equation 2
When $x_1 = 1$ (op = 3) Equation 1 Equation 2 Equation 3	7. pass = 1 Equation 2 Equation 3	8. pass = 2 Equation 1 Equation 3	9. pass = 3 Equation 1 Equation 2

After all the eigenvectors have been computed, they are combined into a single matrix V. If V is invertible, we can compute $V^{-1}AV$ to see how good the eigenvectors are. If V^{-1} cannot be computed, there is another search strategy available. The subsets can be searched in reverse order. That strategy is shown in the following table, using **irow = 2**.

Summary of the Exhaustive Search When the Matrix Size Is Three and *irow* = 2			
Base Equations	Subset 1	Subset 2	Subset 3
When $x_3 = 1$ (op = 1) Equation 1 Equation 2 Equation 3	1. pass = 1 Equation 1 Equation 2	2. pass = 2 Equation 1 Equation 3	3. pass = 3 Equation 2 Equation 3
When $x_2 = 1$ (op = 2) Equation 1 Equation 2 Equation 3	4. pass = 1 Equation 1 Equation 2	5. pass = 2 Equation 1 Equation 3	6. pass = 3 Equation 2 Equation 3
When $x_1 = 1$ (op = 3) Equation 1 Equation 2 Equation 3	7. pass = 1 Equation 1 Equation 2	8. pass = 2 Equation 1 Equation 3	9. pass = 3 Equation 2 Equation 3

The flowchart of the program appears on the next page. It shows the following:

- The inner loop, which uses the **pass** flag
- The middle loop, which uses the **op** flag
- The outer loop, which uses the **irow** flag

The program will not cover the case wherein the values of two or more variables must be picked in order to find a solution. The literature calls this a defective matrix (see example 16.3).

Eigenvectors

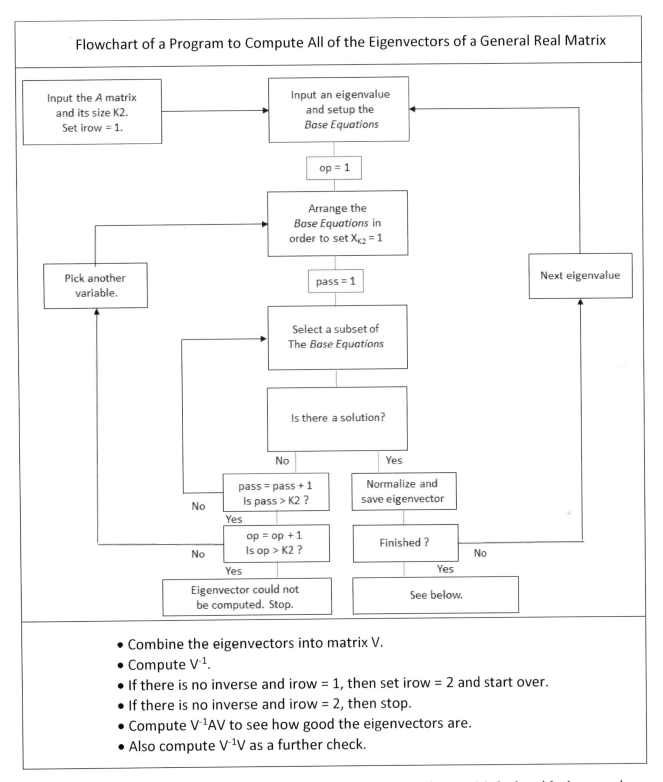

Chapters 4–13 are a programming manual for Excel with VBA. Chapter 14 deals with the actual programs to compute eigenvalues and eigenvectors.

Chapter 4: The Programming Environment between Excel and VBA

Open Excel, and create a new workbook. Then *Save As*:

To facilitate this discussion, enter the following:
Filename: First   Where: Desktop   Type: Excel Macro-Enabled Workbook (.xlsm)
When the file named First is reopened, *click* the Enable Macros button if required.

On a tab near the bottom-left of the screen, the spreadsheet is called **Sheet1**. The + next to this tab is for adding more spreadsheets. For now, leave it at **Sheet1**.

The rows of the spreadsheets are given numbers. The columns are given letters. If you want to change the letters to numbers, see the following table.

How to Switch the Columns from Letters to Numbers
On a PC with Excel 2013: File → Options → Formulas → R1C1 Reference Style.
On a Mac with Excel 2011: Excel → Preferences → General → R1C1 Reference Style.
If these instructions don't work on your system, Google   *How to switch to R1C1 reference style in Excel (your version)*.

If the **Developer** tab is not shown, it must be activated. See the following table.

How to Activate the Developer Tab
On a PC with Excel 2013: File → Options → Customize Ribbon → Developer.
On a Mac with Excel 2011: Excel → Preferences → Ribbon → Developer.
If these instructions don't work on your system, Google   *How to activate the developer tab in Excel (your version)*.

The following is a step-by-step discussion on how to write a **VBA** program.

Click the **Developer** tab. On its ribbon, click **Editor** (or **Visual Basic**). Two things happen.

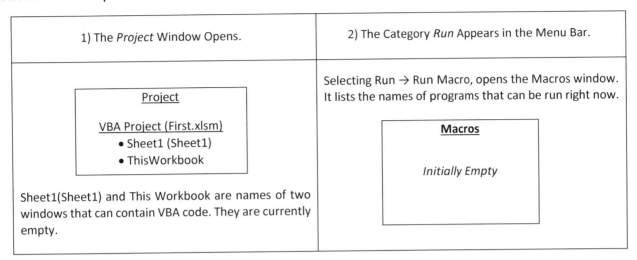

1) The *Project* Window Opens.	2) The Category *Run* Appears in the Menu Bar.
Sheet1(Sheet1) and This Workbook are names of two windows that can contain VBA code. They are currently empty.	Selecting Run → Run Macro, opens the Macros window. It lists the names of programs that can be run right now.

We will now write a program in the **Code** window called **ThisWorkbook**.

1) In the **Project** window, double-click **ThisWorkbook**. The code window will open.

2) At the cursor in the empty window, enter the following six lines of VBA code.

First.xlsm: ThisWorkbook (Code)	
Code	Comments

Sub one()
  x = 1
  Cells(1, 1) = 4 * Atn(x)
  pie = Cells(1, 1)
  Cells(1, 3) = 2 * pie
End Sub
	•The name of the program is *one*. The empty parentheses are necessary. More in chapter 10.
	
	•The *Cells...* statement writes the value of $\pi = 4*\tan^{-1}(1)$ into spreadsheet cell row 1, column 1.
	
	•The *pie=...* statement reads the value of π from cell row 1, column 1.
	
	•This *Cells...* statement writes the value of $2*\pi$ into cell row 1, column 3.

3) In the top menu bar, select File→Save, and close the window.

4) Select **Run→Run Macro**. This opens the **Macros** window. It will show an item called **ThisWorkbook.one**. Click on this item, and then click **Run**.

5) On the spreadsheet, see the following:

- The value of π is written in the cell row 1, column 1.
- The value of 2π is written in row 1, column 3.

6) Prepare for the next run. Clear the spreadsheet by selecting (highlighting) the data, and then click **Edit→Clear→All**. Selecting is done by holding down the clicker and dragging.

Writing Another Program

1) Via the + at the bottom of the spreadsheet, add **Sheet2**. Then click **Editor (Visual Basic)**. This opens the **Project** window. Observe that **Sheet2(Sheet2)** has been added. The windows **Sheet1(Sheet1)** and **Sheet2(Sheet2)** are currently empty.

Project
VBA Project (First.xlsm)
Sheet1(Sheet1)
Sheet2(Sheet2)
ThisWorkbook

2) Select **Run→Run Macro**. In the **Macros** window, click **ThisWorkbook.one** and then **Run**.

3) This time, the values of π and 2π are written on **Sheet2**. This shows that the program named **one**, in **ThisWorkbook**, reads from and writes to the open (visible) spreadsheet.

Writing More Programs

We will now add another code window. If the **Project** window is not open, click **Editor**. Then select **Insert→Module**. The **Project** window now contains four code windows.

Project
VBA Project (First.xlsm)
Module1
Sheet1(Sheet1)
Sheet2(Sheet2)
ThisWorkbook

We will now copy our program **one** from **ThisWorkbook** to the other three code windows.

- Double-click **ThisWorkbook**, select **Edit→Select All→Copy**, and close the window.

- Double-click **Module1**, select **Edit→Paste** and then **File→Save**, and close the window.

- Double-click **Sheet1**, select **Edit→Paste** and then **File→Save**, and close the window.

- Double-click **Sheet2**, select **Edit→Paste** and then **File→Save**, and close the window.

Now, select **Run→Run Macro** to open the **Macros** window.

Macros
one
Sheet1.one
Sheet2.one
ThisWorkbook.one

The following is a discussion of the programs in the Macros window.

- ThisWorkbook.one communicates with the open (visible) spreadsheet.

- The program one that is in Module1 also communicates with the open spreadsheet.

- Sheet1.one will communicate only with Sheet1.

- Sheet2.one will communicate only with Sheet2.

Summary and Notes

- The code window ThisWorkbook communicates with the open sheet.

- The code windows Module$j$ communicate with the open sheet.

- The code windows Sheet$j$ communicate only with Sheet$j$.

- By using ThisWorkbook and the Module$j$ code windows, a program can run many cases, changing only the spreadsheets between cases.

- All programs in this book use the read-write statements that have been shown thus far. There is another way, which will be defined at the end of this chapter.

- Code windows can contain more than one program.

- Programs in different code windows communicate with one another through the spreadsheets.

- Programs in the same window can also communicate with one another through the spreadsheets. But as will be shown in chapter 10, they can also communicate with one another directly.

Printing the Spreadsheet Data

Spreadsheet data can be printed directly from Excel. Data can also be pasted into Word and Powerpoint. In that way, they can be further annotated.

More about Code Windows and Spreadsheets

- Programs read from and write to specific cells. Cutting and pasting data from these cells to others can keep the specific cells free for subsequent runs.

- An attractive feature of the spreadsheet is that, after a run, clarifying comments can be manually added.

- To select two or more areas on the spreadsheet, do the following:
  - First area: hold down the clicker and drag.
  - Subsequent areas: hold down the command key while holding down the clicker and dragging, or hold down the ctrl key while holding down the clicker and dragging.

- All code windows can be given usernames. The Properties window is where this is done. To access this window from the menu bar, select View→Properties Window.

- To view the code, go to View→Code. This opens a code window just like double-clicking on its name.

- The Project window contains all the files that are open. This facilitates things like copying from one file to another.

- It is good practice to close code windows before selecting Run→Run Macro.

Other Read-Write Statements

These statements allow any program in any window to read from and write to any spreadsheet. Since they are not used in any program in this book, the reader may skip to chapter 5.

The following table shows two programs in the Sheet1 code window. Program one is repeated. Program two uses these alternate read-write statements.

First.xlsm: Sheet1 (Code)	
Code	Comments
| ```
Sub one()
    x = 1
    Cells(1, 1) = 4 * Atn(x)
    pie = Cells(1, 1)
    Cells(1, 3) = 2 * pie
End Sub

Sub two()
    x = 1
    Worksheets("Sheet4").Cells(1, 1) = 4 * Atn(x)
    pie = Worksheets("Sheet4").Cells(1, 1)
    Worksheets("Sheet5").Cells(1, 3) = 2 * pie
    Cells(1, 5) = 2 * pie
End Sub
``` | •Sub one reads and writes only to Sheet1.<br><br>•Sub two reads and writes to Sheet4 and Sheet5.<br><br>*Note:* Sub two also writes to Sheet1 by default. |
| The prefix Worksheets("Sheet *j* ") directs the Cells statement to Sheet *j*, only if Sheet *j* has been added to the Project Window. ||
```

Chapter 5: Making Graphs Using VBA

This chapter describes three programs and how to graph their outputs.

Program graph1

This program reads values of y from a spreadsheet, computes y^2 and y^3, and then writes these back onto the same spreadsheet. This involves repeating a series of calculations a predetermined number of times. It will use the VBA statements for looping known as For...Next.

Syntax for Looping via For...Next
For index = start-value To end-value Step step-value   {statements in the loop} Next index

Notes:
- Index is a variable (integer or floating point).
- Start-value, end-value, and step-value may be integers or floating points and positive or negative.
- If Step is not used, Step 1 is the default.

The following is a listing of the program and its output.

graph1.xlsm: ThisWorkbook (Code)

```
Sub graph1()
For i = 1 To 9
    y = Cells( i, 1 )
    Cells( i, 2 ) = y
    Cells( i, 3 ) = y ^ 2
    Cells( i, 4 ) = y ^ 3
Next i
End Sub
```

The Spreadsheet from Program graph1

Y (user input) Column 1	Y (series 1) Column 2	Y^2 (series 2) Column 3	Y^3 (series 3) Column 4
-1.5	-1.5	2.25	-3.375
-1.2	-1.2	1.44	-1.728
-0.8	-0.8	0.64	-0.512
-0.4	-0.4	0.16	-0.064
0	0	0	0
0.4	0.4	0.16	0.064
0.8	0.8	0.64	0.512
1.2	1.2	1.44	1.728
1.5	1.5	2.25	3.375

Values of y are input by the user in column 1. The program reads these values and then writes them into column 2. This column will be called **Series 1**. For each value of y, the program then computes y^2 and y^3 and writes these into columns 3 and 4.

The following is a plot of the output and how the plot was made.

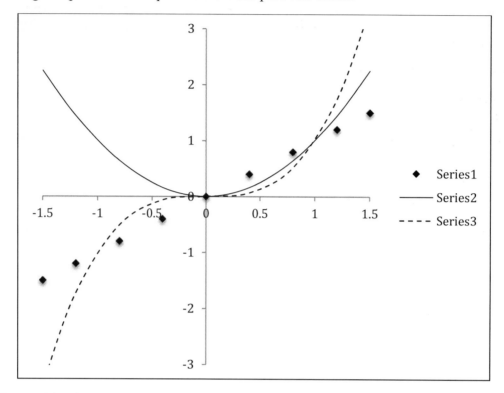

Excel refers to this plot as a **Scatter** chart or X-Y chart. Of the columns of data selected to be plotted, the horizontal axis is on the left. All the other columns are for the vertical axis, and they are called **Series 1, 2**, and so on. All the series share the horizontal axis. Each series has its own format, for example, solid line, dashed line, and so on.

If the points are to be connected, the horizontal-axis data must be monotonic. Each axis has its own scaling (minimum, maximum, increment). The scales for either or both axes can be logarithmic.

The following two tables show how to make a **Scatter** chart. One is for the PC, and the other is for the Mac.

Making Graphs

One Way to Make Scatter Charts on a PC with Excel 2013

1: Select (highlight) the data to be plotted. For Program graph1, this area is from cell(1,1) to cell(9,4).

2: Select the Insert Tab. In the chart area of the ribbon, insert a Scatter chart with Smooth Lines.

3: Click on the chart. This adds chart tool tabs Design and Format. Select Design. On this ribbon, select Add Chart Elements. Select Chart and Axis Titles, which adds *text boxes*. Select Gridlines.

4: Click on the chart, and select the Format Tab. On the chart itself, do the following:

- Double-click on the vertical axis. In the *pop-up window*, select the *three-bars icon* for Axis Options. Enter range, increments, horizontal axis crossing, and maybe log scale. Repeat for horizontal axis.

- Double-click on series 1 plot (or 2, etc.). In the *pop-up window*, select *paint bucket icon* for Fill & Line.

    - If the plot is a *line*, select Line and then color, dash type, and width.

    - If the plot is *points*, select Marker and then Marker Options for type and size and Fill for color.

    - *Note:* To change a series from a *line* to *points* (or vice versa), select the series. Then again, select Insert Tab, Scatter chart, and this time, the Scatter Points icon.

5: Size the chart.
- Select the Format Tab and on this ribbon, type in the size (height and width).
or, • With the *cursor* on the borders of the chart, click and drag the chart area adjusters.

One Way to Make Scatter Charts on a Mac with Excel 2011

1: Select (highlight) the data to be plotted. For Program graph1, this area is from cell(1,1) to cell(9,4).

2: Select the Charts Tab. On this ribbon, insert a Scatter chart with Smooth Lined Scatter.

3: Select Chart Layout tab. Select Chart and Axis Titles to add *text boxes*. Select Gridlines.

4: Click on the chart, and select the Format Tab. On the chart itself, do the following:

- Double-click on the vertical axis. In the *pop-up window*, select Scale. Enter the range, increments, and horizontal axis crossing. Note the option for log scale. Repeat for the horizontal axis.

- Double-click on the series 1 plot (or 2, etc.). In the *pop-up window*:

    - If the plot is a *line*, select Line. Then Solid for color and Weights & Arrows for dash and width.

    - If the plot is *points*, select Marker Fill and Solid for color and Marker Style for style and size.

    - *Note:* To change a series from a *line* to *points* (or vice versa), select the series. Then again, select the Charts Tab, Scatter, and this time, Marked Scatter.

5: Size the chart.
- Select the Format Tab, and on this ribbon, type in the size (height and width).
or, • With the *cursor* on the borders of the chart, click and drag the chart area adjusters.

Program graph2

This program computes e^x and $\ln(x)$ as x ranges from -2 to 3. This program shows how to plot several series when each has different horizontal-axis values. The program introduces the following VBA statements.

VBA Function	Syntax
$y = e^x$	y = Exp(x)
$y = \ln(x)$	y = Log(x)
y = x rounded to a specified number of decimal places	y = Application.Round(x, nd)
y = a list of numbers assigned to y	y = Array(number 1, number 2, …)
y(1) refers to the first number in the array	Option Base 1
y(0) refers to the first number in the array	Option Base 0 (default value)

The following is the program code.

graph2.xlsm: ThisWorkbook (Code)	

`Option Base 1`
`Sub graph2()`
 	
`Xa = Array( -2, -1.5, -1, -0.75, -0.5, -0.25, 0, 0.6, 1.25)`
`For i = 1 to 9`
`  Y = Exp( Xa( i ) )`
`  Cells( i, 1) = Xa( i )`
`  Cells( i, 2 ) = Application.Round( Y, 2 )`
`Next i`
 	
`Xb = Array( 0.01, 0.05, 0.1, 0.2, 0.5, 0.75, 1, 1.25)`
`For i = 1 to 8`
`  Y = Log( Xb( i ) )`
`  Cells( i + 9, 1) = Xb( i )`
`  Cells( i + 9, 3 ) = Application.Round( Y, 2 )`
`Next i`
 	
`Xc = Array( 0.5, 1, 2, 3, 1.5, 2.5)`
`For i = 1 to 6`
`  Y = Log( Xc( i ) )`
`  Cells( i + 9 + 8, 1) = Xc( i )`
`  Cells( i + 9 + 8, 4 ) = Application.Round( Y, 2 )`
`Next i`
 	
`End Sub`
	Option Base 1 allows Xa(1) = -2 (see above table or chapter 6).
	
	The function e^x is computed from the numbers in the Xa array. Since this is a list of monotonic numbers with uneven intervals, it can be formatted with a dashed line. It is printed in column 2 and plotted as *Series1*.
	
	The function y = ln(x) is computed from the numbers in the Xb array. Since it is a list of monotonic numbers with uneven intervals, it can be formatted with a solid line. It is printed in column 3 and plotted as *Series2*.
	
	The function y = ln(x) is computed from the Xc array. This is a list of non-monotonic numbers with uneven intervals. Hence it must be formatted with points. It is printed in column 4 and plotted as *Series3*.
	
	The arrays Xa, Xb, and Xc are printed consecutively in column 1, in the rows that correspond to their individual functions.
	
	Note that arithmetic can be done in the *Cells* statements.

The following is the output.

The Spreadsheet from Program graph2				
	x axis Column 1	$y = e^{X_a}$ Column 2	$y = \ln(X_b)$ Column 3	$y = \ln(X_c)$ Column 4
X_a	-2.00	0.14		
	-1.50	0.22		
	-1.00	0.37		
	-0.75	0.47		
	-0.50	0.61		
	-0.25	0.78		
	0	1.00		
	0.60	1.82		
	1.25	3.49		
X_b	0.01		-4.61	
	0.05		-3.00	
	0.10		-2.30	
	0.20		-1.61	
	0.50		-0.69	
	0.75		-0.29	
	1.00		0	
	1.25		0.22	
X_c	0.50			-0.69
	1.00			0
	2.00			0.69
	3.00			1.10
	1.50			0.41
	2.50			0.92

The following is the plot. The data selected for the plot is the area from cell(1, 1) to cell(23, 4).

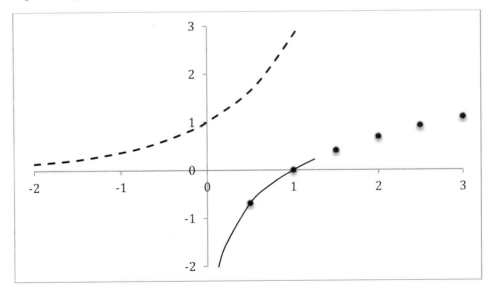

On **Scatter** charts, an empty cell is like *lifting the pen while writing*.

Program graph3

Excel has another kind of chart called a **Surface** chart. It is like a topographical map. Its three axes are vertical, horizontal, and depth. The vertical-axis data has numerical values, whereas the data for the other axes are bins or categories. But these categories can be numbers. If they correspond to actual values, the **Surface** chart becomes a true 3-D chart. This is demonstrated in the following program.

This program computes $f(i,j) = \sqrt{(i-2)^2 + (j-2)^2}$, for $i = 0$ *to* 4 and $j = 0$ *to* 4. The values of $f(i,j)$ are output to the spreadsheet in matrix form to be used for the vertical axis. The values for i and j are the categories for the depth and horizontal axes.

The following is a listing of the code and the spreadsheet output. The code illustrates the use of nested **For...Next** loops as well as the syntax for computing the square root.

graph3.xlsm: ThisWorkbook (Code)

```
Sub graph3()
For i = 0 To 4
    For j = 0 To 4
        f = Sqr( ( i – 2 )² + ( j – 2 )² )
        Cells( 2 + i, 2 + j ) = Application.Round( f, 2 )
        Cells ( 1, 2 + j ) = j
    Next j
    Cells( 2 + i, 1 ) = i
Next i
End Sub
```

The Spreadsheet from Program graph3

		Values of *j* Used as Horizontal Axis Categories				
	Column 1	Column 2	Column 3	Column 4	Column 5	Column 6
Row 1		0	1	2	3	4
Values of *i* Used as Depth Axis Categories	0	2.83	2.24	2	2.24	2.83
	1	2.24	1.41	1	1.41	2.24
	2	2.00	1.00	0	1.00	2.00
	3	2.24	1.41	1	1.41	2.24
	4	2.83	2.24	2	2.24	2.83

The following is a **Surface** chart of these data. The spreadsheet area highlighted for plotting is from cell(1, 1) to cell(6, 6). Note that cell(1, 1) is intentionally blank.

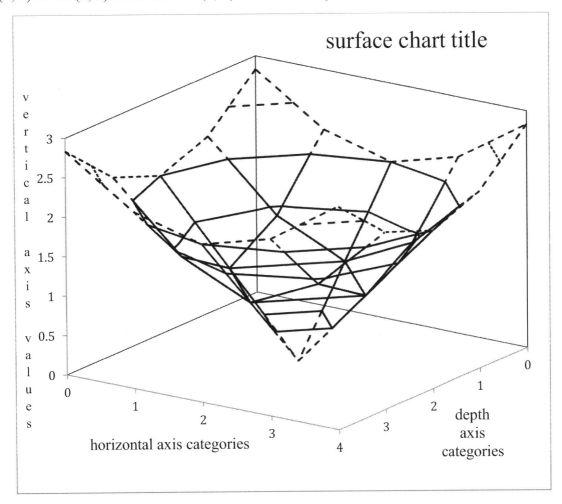

The following two tables show how to make a **Surface** chart. One is for the PC. The other is for the Mac.

Making Graphs

One Way to Make Surface Charts on a PC with Excel 2013

1: Highlight the rectangular area of data to be plotted. For Program graph3, from blank cell(1,1) to cell(9,4).

2: Select the Insert Tab. On this ribbon, insert Surface chart with a Wireframe 3-D Surface.

3: Click on the chart. Select Design Tab and Add Chart Elements. Select Titles for the chart and for the horizontal, vertical, and depth axes. Select Gridlines for these axes.

4: Click on the chart, and select the Format Tab. Then on the chart itself, do the following:

- Double-click on the vertical axis. In the *pop-up window*, select the *three-bars icon* for Axis Options. Enter range, increments, and Floor Crosses At.

- Double-click only Legend Entry 1 (or 2, etc.). In the *pop-up window*, select the *paint bucket icon* for Fill and Line. Then select Border for color, width, and dash.

- Double-click on the depth axis. In the *pop-up window*, select the *three-bars icon* for Axis Options. Then select Axis Options for *series in reverse order*.

- Double-click on the plot area of the chart. In the *pop-up window*, select the *pentagon icon* for Effects. Then 3-D rotation.

5: Size the chart. Same as Scatter chart.

One Way to Make Surface Charts on a Mac with Excel 2011

1: Highlight the rectangular area of data to be plotted. For Program graph3, from blank cell(1,1) to cell(9,4).

2: Select the Charts Tab. On this ribbon, select Other and Wireframe 3-D Surface.

3: Click on the chart. Select the Chart Layout Tab. Select Titles for the chart and for the horizontal, vertical, and depth axes. Select Gridlines for these axes.

4: Click on the chart, and select the Format Tab. Then on the chart itself, do the following:

- Double-click on the vertical axis. In the *pop-up window*, select Scale, and then enter range, increments, and Floor Crosses At.

- Double-click only Legend Entry 1 (or 2, etc.). In the *pop-up window*, select Line and Solid for color and Weights & Arrows for width and dash.

- Double-click on the depth axis. In the *pop-up window*, select Scale, and then maybe select *series in reverse order*.

- Double-click on the plot area of the chart. In the *pop-up window*, select 3-D Rotation.

5: Size the chart. Same as Scatter chart.

Printing a Chart

A chart can be printed directly from **Excel**. It can also be pasted into **Word** and **PowerPoint**. One way to paste a chart into **Word** without the border is shown in the following table.

> - Click on the chart and copy.
> - Paste into PowerPoint and resize, reformat, and annotate if desired.
> - Copy from PowerPoint.
> - Paste into Word via Paste Special → PDF or Paste Special → Bitmap.
> - *Note:* The use of PowerPoint removes the border.

Another way to paste into **Word** is:

$$\text{paste} \to \text{paste special} \to \text{MicrosoftOfficeGraphicObject}.$$

This allows you to alter the appearance of the chart right in **Word**.

Chapter 6: VBA Arrays and Data Types

Variables that are assigned multiple values are called *arrays*. *Indices* are used to refer to a particular value in an array. For example, if five values are assigned to variable X, then X appears in code as X(i), and i can range from 1 to 5. Then the statement Y = X(3) assigns to Y the value in the X array for i = 3.

Array variables can have more than one index. Consider the variable beta(i, j, k). If i ranges from 1 to 2, j ranges from 1 to 4, and k ranges from 1 to 5, then beta can have 40 values.

The Dim Statement

The program compiler (interpreter) must know in advance how much space in memory to set aside for each array variable. This is done via the Dim statement. All arrays must be *sized* in a Dim statement. The one exception will be discussed later. Here is an example of a Dim statement.

$$\text{Dim u(9,10),v(5)}$$

This statement allocates 90 spaces for u and 5 spaces for v.

- The Dim statement can allocate more spaces than are actually needed.

- There may be multiple Dim statements.

- A Dim statement can be placed anywhere in the code, as long as it appears before each of its variables is used.

Another Use of the Dim Statement

When the VBA statement Option Explicit is the first line of the code, all variables used in the program must appear in a Dim statement. With this statement in the code, the names of all the variables are easily seen in the Dim statements.

A Third Use of the Dim Statement

All variables in a program are classified according to their *types*. *Types* include integer, floating point, character, logical, and so on. When it is necessary to declare that a variable is a certain type, this is done in a Dim statement.

VBA has a so-called *smart* type. In fact, it's the default type for all the variables. It is illustrated in the following two programs.

Arrays and Data Types

Program Illustrating the Variant Data Type	
Notes: • If the type is not specified, type *variant* is assumed. Hence, in this program, variable b is a *variant*. • The data type of a variant automatically adapts during the program. • *Variant* is the only data type that is used in any program in this book.	
Code	Comments

Sub variant_1()
   b = " first "
     Cells(1, 1) = b
   b = 11
     Cells(2, 1) = b
   b = 77.5
     Cells(3, 1) = b
   b = Array(" second ", 33, " third ", 3.14159)
     For i = 0 To 3
       Cells(4, i + 1) = b(i)
     Next i
   b = 55
     Cells(5, 1) = b
 End Sub
	• b starts as a character variable whose value is specified within quotes.
	 • Then, b automatically becomes an integer.
	
	 • b becomes a floating point number.
	
	 • b becomes an array of dimension *four*.
	   • By default [Option Base 0], first item is b(0).
	   • When *I* = 0, Cells(4, *I* + 1) = Cells(4, 1).
	
	 • b becomes a floating point or integer number. (The array function has been overridden.)

The following table shows its output.

The Values of b Output by Program variant_1			
Row	Column 1	Column 2	Column 3	Column 4
1	first			
2	11			
3	77.5			
4	second	33	third	3.14159
5	55			

The following program uses the variable b as an array. It produces the same output.

Second Program Illustrating the Variant Data Type	

Sub variant_2()
 Dim b(10)
   b(0) = " first "
     Cells(1, 1) = b(0)
   b(0) = 11
     Cells(2, 1) = b(0)
   b(0) = 77.5
     Cells(3, 1) = b(0)
   b(0) = "second": b(1) = 33: b(2) = "third": b(3) = 3.14159
     For I = 0 To 3
       Cells(4, I + 1) = b(i)
     Next i
   b(0) = 55
     Cells(5, 1) = b(0)
 End Sub
	• The Dim statement has been added.
	
	 • b(0) starts as a character variable.
	
	 • b(0) automatically becomes an integer.
	
	 • b(0) becomes a floating point number.
	
	 • Data types in the b array can differ.
	
	 Note: Multiple statements on a line are separated by colons.
	
	 • b(0) becomes a floating point or integer number.

Option Base

VBA has another option statement called **Base**. When the statement **Option Base 1** is at the top of the code, the indices for all arrays start at unity. If **Option Base 0** is at the top, the indices for all arrays start at zero. **Option Base 0** is the default. The following two programs illustrate this. They produce the same result.

Program Sub one	Program Sub two

Sub one()
alpha = Array(-25, 25, "four", 0, -10)
  For i = 0 To 4
    Cells(1, i + 1) = alpha(i)
  Next i
End Sub
	Option Base 1
	Sub two()
	alpha = Array(-25, 25, "four", 0, -10)
	  For i = 1 To 5
	    Cells(1, i) = alpha(i)
	  Next i
	End Sub

Note the *arithmetic* required in the Cells statement. When i = 0, Cells(1, i + 1) = Cells(1, 1)	

Final Notes

- This is the exception in the Dim statement that was mentioned earlier. A variable given values by the **Array** function may be declared in a **Dim** statement, but it must not be dimensioned in a **Dim** statement. In other words, if X = Array(2,4,6,8), it is permissible to have the statement **Dim X**. It is not permissible to have the statement **Dim X(4)**.

- The **Array** function is a convenient way of assigning values to a one-dimensional variable. It is awkward to use for a variable with two or more dimensions. Also, it can only assign values in the program code. Values cannot be assigned from the spreadsheet.

- The storage space for a **variant** variable is twice that for a double-precision variable. However, its numeric range is the same as that of double precision. From Lomax, use of the **variant** data type does slightly increase program run time.[7]

- All of the variables in the programs in this book are **variant** type.

- When an option statement (**Explicit** or **Base**) is used, it applies to all the programs in that code window.

---

[7] P. Lomax, *VB and VBA in a Nutshell* (O'Reilly, 1998).

Chapter 7: Solving Linear Algebraic Equations Using Matrices and Dynamic Arrays

There are several Excel functions for matrices or rectangular arrays. I demonstrate them while solving linear algebraic equations. Two of the functions are the following:

- Minverse is used to invert a real matrix.
- MMult is used to multiply two real matrices.

The rules for using these are the following:

- The arrays that are input to Minverse and MMult must be dimensioned to their exact size.

- The arrays that are output by Minverse and MMult must not be sized in a Dim statement. This is like the Array function (see "Final Notes" in chapter 6).

Dynamic Arrays

The Dim statement sets the dimensions of arrays. Functions like Minverse must know the dimensions exactly. The ReDim statement allows the dimensions to be established while the program is running. The following table demonstrates the ReDim statement.

Syntax for the ReDim Statement	
• • Dim A( ) • • •	• The empty parentheses declares A to be a dynamic array.
• n=3 ReDim A(n, n) • •	• Later in the program, the size of the A matrix becomes known. • This statement dimensions the A matrix.
Notes: • The statement Dim A(n, n) is not allowed. • The ReDim statement can be used only in the same program wherein its Dim statement appears.	

The following two programs illustrate the use of Minverse, MMult, and the ReDim statements. Note the use of Option Base 1.

Program inverse

This program inverts the 6 × 6 Hilbert matrix. Each of its i and j cells are formed by the following equation:

$$\frac{1}{(i + j - 1)}$$

Each cell of the inverse of a Hilbert matrix is an integer.

Program inverse	
Code	Comments

Option Base 1
Sub inverse()
Dim A()
  n = 6
  ReDim A(n, n)
  For i = 1 To n
    For j = 1 To n
      A(i, j) = 1 / (i + j - 1)
      Cells(i, j) = Application.Round(A(i, j), 6)
    Next j
  Next i
  Ainv = Application.MInverse(A)
  ident = Application.MMult(Ainv, A)
  For i = 1 To n
    For j = 1 To n
      Cells(i, j + n) = Ainv(i, j)
      Cells(i + n, j) = Application.Round(ident(i, j), 1)
    Next j
  Next i
End Sub
	• Declare A to be a dynamic matrix.
	• n is the size of the A matrix.
	•The A matrix is sized exactly (required by Minverse).
	
	• Compute and print out the A matrix.
	
	• Ainv = A^{-1}
	• ident = $A^{-1}*A$ (This provides an accuracy check.)
	
	•Print out the results.

The following table shows the output.

The Output of Program inverse											
Hilbert Matrix (n = 6)						Hilbert Matrix Inverse					
1	1/2	1/3	1/4	1/5	1/6	36	-630	3360	-7560	7560	-2772
1/2	1/3	1/4	1/5	1/6	1/7	-630	14700	-88200	211680	-220500	83160
1/3	1/4	1/5	1/6	1/7	1/8	3360	-88200	564480	-1411200	1512000	-582120
1/4	1/5	1/6	1/7	1/8	1/9	-7560	211680	-1411200	3628800	-3969000	1552320
1/5	1/6	1/7	1/8	1/9	1/10	7560	-220500	1512000	-3969000	4410000	-1746360
1/6	1/7	1/8	1/9	1/10	1/11	-2772	83160	-582120	1552320	-1746360	698544

- On the actual spreadsheet, the Hilbert matrix is printed using decimal points.

- Program inverse will run without the ReDim statement if Dim A() is replaced by Dim A(6,6).

- The identity matrix is not shown, but it was indeed the identity matrix.

Program lineq1

This program solves the following linear algebraic equation:

$$A * x = b \quad \text{where} \quad A = \begin{bmatrix} 1 & -1 & 3 \\ 1 & 2 & -2 \\ 3 & -1 & 5 \end{bmatrix} \quad \text{and} \quad b = \begin{bmatrix} 4 \\ 10 \\ 14 \end{bmatrix}.$$

The program finds x from this equation:

$$x = A^{-1} * b.$$

The matrices and their sizes are input from the spreadsheet.

Program lineq1	
Code	Comments

Option Base 1
 Sub lineq1()
 Dim A(), b()
 n = Cells(1, 1)
 ReDim A(n, n), b(n, 1)
 For i = 1 To n
   For j = 1 To n
     A(i, j) = Cells(i, j + 1)
   Next j
 Next i
 For i = 1 To n
   b(i, 1) = Cells(i, n + 2)
 Next i
 Ainv = Application.MInverse(A)
 	
 x = Application.MMult(Ainv, b)
 	
 bcheck = Application.MMult(A, x)
 For i = 1 To n
   Cells(i, 6) = x(i, 1)
   Cells(i, 7) = bcheck(i, 1)
 Next i
 End Sub
	• Declare A and b to be dynamic matrices.
	 • n is the size of the A matrix.
	 • A and b can now be sized exactly (b *is sized as a column matrix*).
	
	 • A and b are read from the spreadsheet.
	
	 • Ainv = A^{-1}
	
	 • x = A^{-1}*b
	
	 • bcheck = A*x This checks the answer.
	
	 • x and bcheck are printed to the spreadsheet.

The following table shows the input and output.

The Input and Output of Program lineq1						
Input					Output	
n	A(i, j)			b(i, 1)	x(i, 1)	bcheck(i, 1)
3	1	-1	3	4	2	4
	1	2	-2	10	7	10
	3	-1	5	14	3	14

Program lineq2

This program solves the following linear algebraic equation:

$$x = A^{-1} * b \quad \text{where} \quad A = \begin{bmatrix} 1 & 1/2 & 1/3 \\ 1/2 & 1/3 & 1/4 \\ 1/3 & 1/4 & 1/5 \end{bmatrix} \quad \text{and} \quad b = \begin{bmatrix} 1 \\ 1 \\ 1 \end{bmatrix}.$$

For this program, the matrix b is specified by the Array function. This makes it a row matrix. It will have to be transposed by the Transpose function. This program also demonstrates the MDeterm function that computes the determinant of a matrix. The matrices and their size are in the program code.

Program lineq2

Code	Comments

Option Base 1
Sub lineq2()
Dim A()
n = 3:   Cells(1, 1) = n
ReDim A(n, n)
For i = 1 To n
  For j = 1 To n
    A(i, j) = 1 / (i + j - 1): Cells(i, j + 1) = A(i, j)
  Next j
Next i
b = Array(1, 1, 1)
For i = 1 To n
                       Cells(i, n + 2) = b(i)
Next i
bT = Application.Transpose(b)
 	
Ainv = Application.MInverse(A)
 	
x = Application.MMult(Ainv, bT)
 	
bcheck = Application.MMult(A, x)
 	
deter = Application.MDeterm(A)
For i = 1 To n
  Cells(i, 6) = x(i, 1): Cells(i, 7) = bcheck(i, 1)
Next i
Cells(1, 8) = deter
End Sub
	• Declare A to be a dynamic matrix.
	• n is the size of A.
	• A can now be dimensioned exactly per the requirements of Minverse.
	
	• Compute and print out the A matrix.
	
	• Set the values of b. Since this is done with the Array function, b is a row matrix.
	
	• Transpose b to allow multiplication by the A^{-1} matrix.
	
	• Ainv = A^{-1}
	
	• x = $A^{-1}*b^T$
	
	• bcheck = A*x
	
	• deter is the determinant of the A matrix.
	
	• Print out **x**, bcheck, and deter.

Program lineq2 Output

n	A			b	x	bcheck	determinant
3	1	1/2	1/3	1	3	1	
	1/2	1/3	1/4	1	-24	1	.000463
	1/3	1/4	1/5	1	30	1	

Chapter 8: Functions

Chapter 7 introduced the functions Minverse, MMult, MDeterm, and Transpose. More functions are defined in this chapter. There are three types of functions:

- Excel functions
- VBA functions
- User-defined functions

All of Excel's functions are accessible to VBA programs. If a VBA function is the same as an Excel function, the VBA function must be used.

- The syntax for an Excel function: y = Application.name(argument list)
- The syntax for a VBA function: y = name(argument list)
- The syntax for a user-defined function: y = name(argument list)

The following four tables list common Excel and VBA functions.

Trig and Math Functions	
$y = \sin(\theta)$ with θ in radians	y = Sin(x)
$y = \cos(\theta)$ with θ in radians	y = Cos(x)
$y = \tan(\theta)$ with θ in radians	y = Tan(x)
$\theta = \sin^{-1}(x)$ with θ in radians	y = Application.Asin(x)
$\theta = \cos^{-1}(x)$ with θ in radians	y = Application.Acos(x)
$\theta = \tan^{-1}(x)$ with θ in radians	y = Atn(x)
$\theta = \tan^{-1}(x2/x1)$ with θ in radians	y = Application.Atan2(x1, x2)
y=square root of x	y = Sqr(x)
$y = e^x$	y = Exp(x)
y=log x to base 10	y = Application.Log10(x)
y=natural log of x	y = Log(x)
y=log x to base a	y = Application.Log(x, a)
y=absolute value of x	y = Abs(x)
y=sign of x (+1, 0 or -1)	y = Sgn(x)

Functions That Place the Decimal Point	
y = integer by truncating x after the decimal point	y = Fix(x)
y = nearest integer to x in negative direction	y = Int(x)
y = nearest integer to x by rounding up or down	y = Round(x)
y = x rounded (up or down) to nd decimal places	y = Application.Round(x, nd)
y = x rounded up to nd decimal places	y = Application.RoundUp(x, nd)
y = x truncated to nd decimal places	y = Application.RoundDown(x, nd)
Note: nd may be positive, zero, or negative.	

Array Functions

- *Note:* For Excel functions, their input arrays must be dimensioned exactly, and the output array must not be dimensioned (see chapter 6).
- By default, all array indices start at 0. To change to 1, use Option Base 1.

y = maximum value of a general array	y = Application.Max(x)
y = minimum value of a general array	y = Application.Min(x)
y = inverse of x matrix	y = Application.Minverse*(x)
y = product of two matrices	y = Application.MMult(x1, x2)
y = determinant of x matrix	y = Application.MDeterm(x)
y = transpose of x matrix	y = Application.Transpose(x)
y = list of numbers assigned to a row array	y = Array(x1, x2, x3,..., xn)

*Minverse returns a scalar when its input is a 1x1 matrix.

Functions

Complex Functions in Excel	
$y = (a + j\,b)$ combined into a single complex number	y = Application.Complex(a, b)
y = real coefficient of a complex number	y = Application.ImReal(x)
y = imaginary coefficient of a complex number	y = Application.Imaginary(x)
y = magnitude of a complex number	y = Application.ImAbs(x)
y = angle (in radians) of a complex number	y = Application.ImArgument(x)
y = quotient of two complex numbers	y = Application.ImDiv(x, z)
y = product of two or more complex numbers	y = Application.ImProduct(x, z, w)
y = sum of two or more complex numbers	y = Application.ImSum(x, z, w)
y = difference of two complex numbers	y = Application.ImSub(x, z)
y = complex number raised to an integer power	y = Application.ImPower(x, n)
$y = e^x$ where x is a complex number	y = Application.ImExp(x)

Functions

The following table shows the syntax for a user-defined function.

Syntax for a User-Defined Function

Sub main()
    •
    •
    y = function_name(argument list)
    •
    •
End Sub
 	
Function function_name(argument list)
    •
    •
    Exit Function   'Optional
    •
    •
    function_name = expression
End Function

The following program illustrates a user-defined function. The function returns n! (n factorial) for a positive integer.

Example *nfactorial*	
Code	Syntax

Sub main()
  n = Cells(1, 1)
  X = nfactorial(n)
  Cells(1, 2) = X
End Sub
 	
Function nfactorial(n)
  nfactorial = 1
  For i = 1 To n
    nfactorial = nfactorial * i
  Next i
End Function
	• If n = 0, the statements in the For...Next loop are not executed. That achieves the result
	
	    0 ! = 1 ! = 1
	
	• The equal sign really means:
	    is replaced by.
	This is shown by the statement:
	    nfactorial = nfactorial * i

Chapter 9:

Looping and Branching
and
The Operators (Comparison, Logical, and Arithmetic)

This chapter is divided into the following sections:

- A: Looping a definite number of times via **For...Next**
- B: Looping an indefinite number of times via **Do...Loop**
- C: Branching according to the *true* or *false* result of a condition test via **If...Then...Else**
- D: Branching by comparing values using the **Select Case** statement
- E: The operators used in the following:
  - Arithmetic (addition, subtraction, etc.)
  - Condition tests (comparison and logical)

Section A: Looping via For...Next

Looping a definite number of times has been introduced in previous chapters. It's appropriate to repeat it here.

Syntax for For...Next

For index = start-value To end-value Step step-value
    {statements in the loop that may include the Exit For statement}
Next index
 	
where: 1. index is an integer or floating point variable.
        2. start-value, end-value, and step-value may be integer or floating point and positive or negative.
        3. Step is optional. Step 1 is the default.
        4. If end-value is less than start-value, execution transfers to after the For...Next loop.
        5. Exit For is a way to exit the loop before completion. It causes the program to branch to the statement after Next index. It is useful for program checkout. Exit For is optional.

Section B: Looping via Do...Loop

The **Do...Loop** statement repeatedly executes a block of code *while* or *until* some condition is satisfied. Section E of this chapter discusses these condition tests. **Do...Loop** has five types.

• **Do While...Loop**: This executes the statements in the loop while some condition is true, and the condition is tested *before* the loop starts.

Syntax for Do While...Loop
Do While { *condition test* }   {statements that may include Exit Do} Loop
Note: • If the *condition* test fails the first time, the statements in the loop will not be executed at all. • Exit Do is optional.

The following program demonstrates the **Do While...Loop**. This program estimates a computer's precision. The program determines the smallest difference between two numbers that the host computer can detect. Starting with an initial value of **epsilon** = 1, the program repeatedly divides **epsilon** by 2 until (1 + epsilon = 1).

The VBA Code for the Machine Epsilon Test Using a Do While...Loop	
Sub epsilon_test() epsilon = 1 Do While (1 + epsilon) > 1 epsilon = epsilon / 2 Loop epsilon = 2 * epsilon End Sub	*Important Note:* This program shows a lesson about using While. Since the test fails inside the Do...Loop, an adjustment must be made outside the loop. That's the reason for the last line of code: epsilon = 2 * epsilon. *Conclusion:* A computer can detect a difference between numbers as small as this value of epsilon. For the Macbook OSX and the Hewlett-Packard 15 with Windows 8.1, epsilon = 2^{-52}.

• **Do...Loop While**: This executes the statements in the loop while some condition is true, but the condition is tested *after* the loop is done.

Syntax for Do...Loop While
Do   {statements that may include Exit Do} Loop While { *condition test* }
Note: • The statements in the loop will be executed at least once. • Exit Do is optional.

The word **While** can be replaced by **Until**. To make this replacement, adjustments must be made to the condition test. This will be shown in a sample program below.

- **Do Until...Loop:** This executes the statements in the loop until some condition is true, where the condition is tested *before* the loop starts.

Syntax for Do Until...Loop

Do Until { *condition test* }
          {statements that may include Exit Do}
      Loop
Note:
• If the *condition test* fails the first time, the statements in the loop will not be executed at all.
• Exit Do is optional.

The following is a program demonstrating the **Do Until...Loop**. This is done by modifying the previous program that used the **Do While...Loop**. The condition test must be changed. The change is seen in the line of code **Do Until (1 + epsilon) = 1**.

The VBA Code for the Machine Epsilon Test using a Do Until...Loop	

Sub epsilon_test()
  epsilon = 1
  Do Until (1 + epsilon) = 1
    epsilon = epsilon / 2
  Loop
  epsilon = 2 * epsilon
End Sub
	Important Note: This program demonstrates a lesson about using Until. Since the *condition test* fails inside the Do...Loop, an adjustment must be made outside the loop. That's the reason for the last line of the code:
	          epsilon = 2 * epsilon

- **Do...Loop Until:** This executes the statements in the loop until some condition is true, but the condition is tested *after* the loop is done.

Syntax for Do...Loop Until

Do
          {statements that may include Exit Do}
      Loop Until { *condition test* }
Note:
• The statements in the loop will be executed at least once.
• Exit Do is optional.

- **Do...Loop** (test not included):

Syntax for Do...Loop

Do
    {statements that must include how the loop is terminated
    and may include Exit Do}
  Loop

Section C: Branching via If...Then...Else

This statement allows branching according to the true or false result of a condition test. Section E of this chapter discusses these condition tests. Five types of If...Then...Else are shown.

- Branching to a single statement:

  > If { *condition* } Then { single statement if condition is true }

- Branching to a block of statements:

  > If { *condition* } Then
  >   { statements if condition is true }
  > End If

- Branching at a fork:

  > If { *condition* } Then
  >   { statements if condition is true }
  > Else
  >   { statements if condition is false }
  > End If

- Branching at a multiple-condition fork:

  > If { *condition 1* } Then
  >   { statements if condition 1 is true }
  > ElseIf { *condition 2* } Then
  >   { statements if condition 1 is false and condition 2 is true }
  > Else
  >   { statements if condition 1 is false and condition 2 is false }
  > End If

- Another type of branching at a multiple-condition fork:

  > If { *condition 1* } Then
  >   { statements if condition 1 is true }
  > ElseIf { *condition 2* } Then
  >   { statements if condition 1 is false and condition 2 is true }
  > End If
  > *Note:* Nothing is done if conditions 1 and 2 are both false.

Notes:

- ElseIf doesn't have its own End If.
- There can be many ElseIfs and Elses.

Section D: Branching via Select Case

This is branching by comparing the value of a variable to *tags* on several cases. For example, when the variable whichcase = 10, the program branches to case 10. The following is the syntax.

Syntax for Select Case

Select Case *whichcase*
  Case *tag1*
    {statements when *whichcase* = *tag1*, after which control transfers to End Select}
    •
    •
  Case *tagn*
    {statements when *whichcase* = *tagn*, after which control transfers to End Select}
  Case Else
    {statements to execute when there is no match}
End Select

Notes: • *whichcase* is a variable whose value is compared to *tagn*. *whichcase* can be a number (integer or floating point, positive or negative), or it can be a character string. • *tagn* is a constant.

In Select Case, VBA proceeds down the list of *tags* until it finds a match. After executing the statements in that block, control transfers to End Select, even though other *tags* farther down the list would also match. Hence, the order and the content of the *tags* are very important.

The following program shows the three forms that the *tags* can have. The program shows that the *tag* can be a number, a range of numbers, or the Is form. The program determines the value of grade from the input value of score.

Three Forms of Tags	
Code	Explanation

Select Case score
  Case 100
    grade = " excellent "
  Case 90 To 99
    grade = " A "
  Case 80 To 89
    grade = " B "
  Case 70 To 79
    grade = " C "
  Case Is >= 60
    grade = " D "
  Case Else
    grade = " F "
End Select
	1) *tag* can be a number (or string).
	
	2) *tag* can be a range of numbers (or strings), from *lower* to *upper*.
	
	3) *tag* can be the Is form and can use any of the *comparison operators* described in Section E of this chapter.

The following table shows more examples for *tag*.

Tag	Value of whichcase
Case 1	1
Case -1	-1
Case 1, 1.5	1 or 1.5 The comma means *or*
Case 1 To 2	(from 1 to 2)
Case 1 To 2, 4 To 5	(from 1 to 2) or (from 4 to 5)
Case Is = 1	1
Case Is = -1	-1
Case Is = 1, Is = 1.5	1 or 1.5
Case Is >= 1, Is <= 2	Any real number. Since this obviously is not intended, care must be taken.
Case Is < 1	< 1 including the negative numbers
Case "aA a "	"aA a " including spaces between the quotes

Section E: Operators (Comparison, Logical, and Arithmetic)

In looping and branching, *condition tests* that evaluate as *true* or *false* determine the paths through the code. These *condition tests* use what are called comparison and logical operators. The arithmetic operators are addition, subtraction, and so on. This section illustrates the syntax for all of these operators as they apply to numbers. Refer to Lomax for how they apply to character strings.[8]

Arithmetic Operators

The following is a list of the standard arithmetic operators.

Operator for...	VBA Code
Addition	+
subtraction and negation	-
multiplication	*
Division	/
exponentiation	^

Refer to Lomax for the syntax of integer division (\) and integer division remainder (Mod).

Comparison Operators

Whereas the result of an arithmetic expression is a number, the result of a comparison expression is *true* or *false*.

Comparison Expression in Words	VBA Syntax	Result of Comparison
3 is greater than 2	3 > 2	true
3 is less than 2	3 < 2	false

The following is a list of all of the VBA comparison operators.

Operator for the Comparison	VBA Syntax
greater than	>
less than	<
greater than or equal to	> =
less than or equal to	< =
equal to	=
not equal to	<>

---

[8] Ibid.

Logical Operators

Like a comparison expression, the result of a logical expression is *true* or *false*. The following table illustrates logical expressions using the logical operator called **And**:

Logical Expression in Words	VBA Syntax	Result of Comparison
3 is greater than 2 And 1 is greater than 0	3 > 2 And 1 > 0	true
3 is greater than 2 And 1 is less than 0	3 > 2 And 1 < 0	false

The following table illustrates four VBA logical operators. In the table, A and B are logicals, that is, they are *true* or *false*.

VBA Syntax	Result
A And B	true if both A and B are true. Otherwise false.
A Or B	true if either or both A and B are true. Otherwise false.
A Xor B	true if A and B are different. Otherwise false.
A Eqv B	true if A and B are the same. Otherwise false.

Another VBA logical operator is called **Not**. Not-false is true. Not-true is false.

The Rules for Operator Precedence

- First, the arithmetic is done. Within the arithmetic itself, there is also a precedence.

> - All exponentiation is done first.
> - All multiplication and division are done next and from left to right (e.g., 4 / 2 * 2 = 4).
> - All addition and subtraction are done last, from left to right.

- Second, the comparisons are done.

- Third, the logical expressions are evaluated in the following order:

Not
And
Or
Xor
Eqv

When the same operator appears multiple times on the same line, evaluation is from left to right.

Example Demonstrating Operator Precedence

Consider the following valid VBA statement:

$$b = \text{Not } 2\wedge 2 + 1 > 4 - 2 * 2 / 3 * 3 \text{ And } 1 < 0 \text{ Or } 3 > 2. \qquad \text{Equation 1}$$

This statement is from bad programming, but it is valid. To evaluate it, the obvious first step is to do the arithmetic. The statement becomes the following:

$$b = \text{Not } 5 > 0 \text{ And } 1 < 0 \text{ Or } 3 > 2.$$

This statement starts with **Not 5**, which by itself doesn't make sense. However, by using parentheses for grouping, we get **b = Not (5 > 0) And (1 < 0) Or (3 > 2)**. Rewritten, it becomes this:

$$b = \textit{false} \text{ And } \textit{false} \text{ Or } \textit{true}.$$

Now comes a dilemma. Evaluating **b** from left to right, this results:

$$b = \textit{false} \text{ Or } \textit{true} = \textit{true}.$$

Evaluating **b** from right to left, this results:

$$b = \textit{false} \text{ And } \textit{true} = \textit{false}.$$

There must be a rule to tell which one. There is! **And** is evaluated before **Or**. Hence, the following is accurate:

$$b = \textit{false} \text{ Or } \textit{true} = \textit{true}.$$

Is that what the programmer wants?

Tip: Expressions within parentheses are evaluated first.

So, rather than memorize rules, the programmer can and should use parentheses to organize the conditions to get the right answer.

For example, b = TRUE if the parentheses in **Equation 1** are as follows:

$$b = \left(\text{Not}\left((2\wedge 2 + 1) > (4 - 2 * (2/3) * 3)\right) \text{ And } 1 < 0\right) \text{ Or } (3 > 2).$$

Also, **b** = FALSE if the parentheses in **Equation 1** are as follows:

$$b = \left(\text{Not}\left((2\wedge 2 + 1) > (4 - 2 * (2/3) * 3)\right)\right) \text{ And } (1 < 0 \text{ Or } 3 > 2).$$

Final Note: The following VBA statements are valid:

$$b = \text{True}$$
$$\text{Cells}(1, 1) = b$$

The logical variable TRUE is printed in Cells(1, 1).

Chapter 10: The Call Statement

So far all of the programs have been written in the structure shown in the following table.

Sub main()   ' *Note* the empty parentheses.
{VBA statements }
End Sub

This chapter shows how a program can be modularized, with a main program that calls a subprogram to perform a task and return the results to the main.

The following table shows the syntax for using the **Call** statement.

Syntax for a Typical Use of the Call Statement

Sub main()
    Dim { arrays used in Sub main }
 	
     { VBA statements }
 	
    Call *name*(argument list)
 	
     {VBA statements }
 	
 End Sub

Sub *name*(argument list)
    Dim { arrays used in this subprogram, excluding those in the argument list }
 	
     { VBA statements }
 	
 End Sub

Notes: • The argument list contains all of the variables that are transferred *to* and *from*.   • There is no limit on the number of Call statements or subprograms.   • Subprograms can Call other subprograms.   • Except for the argument list, values of the variables in a subprogram must be reset or recomputed each time the subprogram is entered.   • The argument list may be empty.   • The calling program and the called program must be in the same Code Window.   • Only the programs with empty parentheses are listed in the Macros Window.

To demonstrate this, I use a program that multiplies two real matrices.

$$[AB] = [A][B].$$

The dimensions are: A is m × L: B is L × n: and AB is m × n. The following table shows this program, wherein the main part reads in the matrices and calls a subprogram to do the multiplication. The result is then returned to the main, which prints it out.

The Call Statement

Program multR to Multiply Real Matrices	
Code page 1	Code page 2
| ```
Sub multR_main()
  Dim A(20, 21), B(21, 22), AB(20, 22)
  m = Cells(1, 1): L = Cells(1, 2): n = Cells(1, 3)
  For i = 1 To m
    For j = 1 To L
      A(i, j) = Cells(i + 1, j)         ' Read A
    Next j
  Next i
  For j = 1 To L
    For k = 1 To n
      B(j, k) = Cells(j + 1, k + L + 1)  ' Read B
    Next k
  Next j
  Call mult(A, B, AB, m, L, n)           ' Call Subprogram
  For i = 1 To m
    For k = 1 To n
      Cells(i + m + 2, k) = AB(i, k)     ' Printout
    Next k
  Next i
End Sub ' multR_main
``` | ```
Sub mult(A, B, AB, m, L, n)
  For i = 1 To m
    For j = 1 To n
      AB(i, j) = 0
      For k = 1 To L
        abk = A(i, k) * B(k, j)
        AB(i, j) = abk + AB(i, j)
      Next k
    Next j
  Next i
End Sub ' mult
``` <br><br>Notes:<br><br>• The matrices are dimensioned in the main program.<br><br>• The main and the subprogram reside in the same code window. |
| Additional Syntax ||
| • An apostrophe causes the rest of the line to be a comment, which is ignored by the program.
• In the subprogram, the statement AB(i, j) = abk + AB(i, j) illustrates that, in VBA, the equal sign means *is replaced by*. ||

The following example uses this program.

Example Call 1: Use Program **multR** to verify that $\begin{bmatrix} 1 & 1/2 \\ 1/2 & 1/3 \end{bmatrix} \begin{bmatrix} -2 \\ 6 \end{bmatrix} = \begin{bmatrix} 1 \\ 1 \end{bmatrix}$. The following table shows the input and output of this program.

Example Call 1 Spreadsheet					
Input					Output
2	2	1			
1	1/2		-2		1
1/2	1/3		6		1

The following table contains a small program that illustrates three important points about the Call statement.

- The variables in the argument list are only addresses in memory. Hence, the subprogram may use different names than the names used in the calling program.

- The values of the variables in the argument list can be changed anywhere, either in the calling program or in the called program.

- If an argument-list variable requires a Dim statement, the Dim statement that covers it must be in the calling program.

Program Illustrating Argument List Variables	

Sub main()
  Dim a(50)
  a(2) = 0
  Call one(a)
  a(2) = a(2) + 3
  Cells(1, 1) = a(2)
End sub
 	
Sub one(b)
  b(2) = b(2) + 3
  Call two(b)
  b(2) = b(2) + 3
End Sub
 	
Sub two(c)
  c(2) = c(2) + 3
End Sub
	• As their values change, a(2) = b(2) = c(2). This is because these variables reside in the same address. The Dim statement covering these must appear in Sub main(), which is the original *calling* program.
	
	• The final answer is a(2) = 12.

Additional Syntax
• Exit Sub This optional statement causes execution to return to the *calling* program, specifically to the statement right after the Call statement. If used in the main program, it causes execution to stop. • End This optional statement causes execution to stop and can be used anywhere.

A program that multiplies complex arrays is shown in the following table. For comparison, the table also repeats the program for real matrices. The real and imaginary parts of a complex number reside in different cells on the spreadsheet. Seeing the programs side by side helps see how this is handled. The functions for complex numbers are defined in chapter 8.

The Call Statement

Program mult to Multiply Complex Matrices	
Code for Program multR	Code for Program mult
| ```
Sub multR_main()
  Dim A(20, 21), B(21, 22), AB(20, 22)
  m = Cells(1, 1): L = Cells(1, 2): n = Cells(1, 3)
  For i = 1 To m
    For j = 1 To L
       A(i, j) = Cells(i + 1, j)
    Next j
  Next i
  For j = 1 To L
    For k = 1 To n
       B(j, k) = Cells(j + 1, k + L + 1)
    Next k
  Next j
  Call mult(A, B, AB, m, L, n)
  For i = 1 To m
    For k = 1 To n
       Cells(i + m + 2, k) = AB(i, k)
    Next k
  Next i
End Sub ' multR_main

Sub mult(A, B, AB, m, L, n)
  For i = 1 To m
    For j = 1 To n
       AB(i, j) = 0
       For k = 1 To L
          abk = A(i, k) * B(k, j)
          AB(i, j) = abk + AB(i, j)
       Next k
    Next j
  Next i
End Sub ' mult
``` | ```
Sub mult_main()
  Dim A(20, 20), b(20, 20), ab(20, 20), m
  m = Cells(1, 1): L = Cells(1, 2): N = Cells(1, 3)
  For i = 1 To m
    For j = 1 To L
       aR = Cells(i + 1, j)
       aI = Cells(i + 1, L + 1 + j)
       A(i, j) = Application.Complex(aR, aI)
    Next j
  Next i
  For i = 1 To L
    For j = 1 To N
       bR = Cells(i + 1, 2 * (L + 1) + j)
       bI = Cells(i + 1, 2 * (L + 1) + N + 1 + j)
       b(i, j) = Application.Complex(bR, bI)
    Next j
  Next i
  Call mult(A, b, ab, m, L, N)
  For i = 1 To m
    For j = 1 To N
       abR = Application.ImReal(ab(i, j))
       abI = Application.Imaginary(ab(i, j))
       Cells(i + 3 + m, j) = Application.Round(abR, 3)
       Cells(i + 3 + m, N + 1 + j) = Application.Round(abI, 3)
    Next j
  Next i
  Cells(m + 3, 1) = " real": Cells(m + 3, N + 2) = " imag"
End Sub ' mult_main

Sub mult(A, b, ab, m, L, N)
  For i = 1 To m
    For j = 1 To N
       ab(i, j) = Application.Complex(0, 0)
       For k = 1 To L
          abk = Application.ImProduct(A(i, k), b(k, j))
          ab(i, j) = Application.ImSum(abk, ab(i, j))
       Next k
    Next j
  Next i
End Sub ' mult
``` |

Example Call 2: Use Program mult to verify that $\begin{bmatrix} 1+j1 & 2-j2 \\ 2+j2 & -1+j3 \end{bmatrix} \begin{bmatrix} -4+j5 \\ 5-j4 \end{bmatrix} = \begin{bmatrix} -7-j17 \\ -11+j21 \end{bmatrix}$.

The following table shows the input and output of this program.

Example Call 2 Spreadsheet									
Input									Output
2	2	1							
1	2		1	-2		-4	5	-7	-17
2	-1		2	3		5	-4	-11	21

This chapter concludes with two programs that further show the use of the Call statement and the complex functions.

- Program lineq that solves linear algebraic equations with complex coefficients
- Program inverse that inverts a matrix with complex elements

Program lineq for Linear Equations with Complex Coefficients

The system of linear equations is written as: $[A][x] = [b]$, where A and b are complex matrices and x is the solution matrix. Program lineq begins by putting A and b into a partitioned matrix: $[A \mid b]$. The solution x will not change if

- a row is multiplied by a nonzero constant;
- a multiple of one row is added to another; or
- rows are interchanged.

Using these row operations, the program changes $[A \mid b]$ to $[\, I \mid x \,]$, where I is the identity matrix. This is like multiplying both sides of the partitioned matrix by A^{-1}.

$$[A \mid b],$$

$$[A^{-1}A \mid A^{-1}b],$$

$$[\, I \mid x \,].$$

Example Call 3: Solve the following problem using row operations.

$$\begin{bmatrix} 1 & 1/2 \\ 1/2 & 1/3 \end{bmatrix} \begin{bmatrix} x_1 \\ x_2 \end{bmatrix} = \begin{bmatrix} 1 \\ 1 \end{bmatrix}$$

The following table shows the process.

Example Call 3			
Step 1 Form the partitioned matrix	1 1/2	1/2 1/3	1 1
Step 2 Multiply row 1 by 1/2 and subtract it from row 2.	1 0	1/2 1/12	1 1/2
Step 3 Multiply row 2 by 6 and subtract it from row 1	1 0	0 1/12	-2 1/2
Step 4 Divide row 1 by 1, and divide row 2 by 1/12	1 0	0 1	-2 6

The result of Step 4 is: $\begin{bmatrix} 1 & 0 \\ 0 & 1 \end{bmatrix} \begin{bmatrix} x_1 \\ x_2 \end{bmatrix} = \begin{bmatrix} -2 \\ 6 \end{bmatrix}$.

By inspection, we find $x_1 = -2$ and $x_2 = 6$.

Example Call 4: Solve the following problem using row operations.

$$\begin{bmatrix} 0 & 1/2 & 1/3 \\ 1/2 & 1/3 & 1/4 \\ 1/3 & 1/4 & 1/5 \end{bmatrix} \begin{bmatrix} x_1 \\ x_2 \\ x_3 \end{bmatrix} = \begin{bmatrix} 1 \\ 1 \\ 1 \end{bmatrix}.$$

The following tables show the process.

Step 1 Form the partitioned matrix	0	1/2	1/3	1
	1/2	1/3	1/4	1
	1/3	1/4	1/5	1

75

Step 2.1 To avoid zero on the diagonal, interchange rows 1 and 2.	1/2	1/3	1/4	1
	0	1/2	1/3	1
	1/3	1/4	1/5	1

Step 2.2 On the matrix from Step 2.1, multiply row 1 by 2/3 and subtract it from row 3.	1/2	1/3	1/4	1
	0	1/2	1/3	1
	0	1/36	1/30	1/3

Step 2.3 Then multiply row 2 by 1/18 and subtract it from row 3.	1/2	1/3	1/4	1
	0	1/2	1/3	1
	0	0	1/67.5	0.278

Step 3.1 Multiply row 3 by 16.88 and subtract it from row 1.	1/2	1/3	0	-3.688
	0	1/2	1/3	1
	0	0	1/67.5	0.278

Step 3.2 Multiply row 3 by 22.5 and subtract it from row 2.	1/2	1/3	0	-3.688
	0	1/2	0	-5.25
	0	0	1/67.5	0.278

Step 3.3 Multiply row 2 by 2/3 and subtract it from row 1.	1/2	0	0	-0.188
	0	1/2	0	-5.25
	0	0	1/67.5	0.278

	1	0	0	-0.375
Step 4 Divide row 1 by 1/2, row 2 by 1/2, and row 3 by 1/67.5.	0	1	0	-10.5
	0	0	1	18.75

By inspection, we find $x_1 = -0.375$, $x_2 = -10.5$, and $x_3 = 18.75$.

The following table shows Program lineq.

- In the program, A is the partitioned matrix.

- The program finishes with statements from Program **mult** to check the answer.

- Because of roundoff, the program puts a tolerance on zero. For example, if $|d| \leq$ tolerance, then $d = 0$, where tolerance might be 1.0e-6.

The Call Statement

Program lineq for Solving Linear Equations with Complex Coefficients

Code page 1	Code page 2
| ```
Sub lineq_main()
  Dim A(50, 51), xd(50), bd(50), ad(50, 51), bdck(50)
  N = Cells(1, 1)
  For i = 1 To N
    For j = 1 To N + 1
      aR = Cells(i + 1, j)
      aI = Cells(i + 1, j + N + 2)
      A(i, j) = Application.Complex(aR, aI)
    Next j
  Next i
  For i = 1 To N
    For j = 1 To N
      ad(i, j) = A(i, j)
    Next j
    bd(i) = A(i, N + 1)
  Next i
  Call lineq(ad, bd, xd, N, fail)
  If fail = 1 Then
    Cells(N + 3, 1) = "no solution"
  Else
    For i = 1 To N
      bdck(i) = Application.Complex(0, 0)
      For j = 1 To N
        adX = Application.ImProduct(ad(i, j), xd(j))
        bdck(i) = Application.ImSum(bdck(i), adX)
      Next j
    Next i
    For i = 1 To N
      xR = Application.ImReal(xd(i))
      xI = Application.Imaginary(xd(i))
      bR = Application.ImReal(bdck(i))
      bI = Application.Imaginary(bdck(i))
      Cells(i + N + 3, 1) = xR
      Cells(i + N + 3, 2) = xI
      Cells(i + N + 3, 4) = bR
      Cells(i + N + 3, 5) = bI
    Next i
    Cells(N + 3, 1) = " X(real)": Cells(N + 3, 2) = " X(imag)"
    Cells(N + 3, 4) = " b(real)": Cells(N + 3, 5) = " b(imag)"
  End If
End Sub  ' lineq_main
``` | ```
Sub lineq(ad, bd, x, N, fail)
  Dim m(50), A(50, 51)
  fail = 0:  tol = 0.000001: zeroC = Application.Complex(0, 0)
  If N = 1 Then
    If Application.ImAbs(ad(1, 1)) < tol Then ad(1, 1) = zeroC
    If Application.ImAbs(ad(1, 1)) = 0 Then
      fail = 1
    Else
      x(1) = Application.ImDiv(bd(1), ad(1, 1))
    End If
    Exit Sub
  End If
  For i = 1 To N                                              ' Step 1
    For j = 1 To N
      A(i, j) = ad(i, j)
    Next j
    A(i, N + 1) = bd(i)
  Next i
  For p = 1 To N - 1 Step 1                                   ' Steo 2
    For i = p To N
      If Application.ImAbs(A(i, p)) < tol Then A(i, p) = zeroC
      If Application.ImAbs(A(i, p)) <> 0 Then Exit For
      If i = N Then
        fail = 1: Exit Sub
      End If
    Next i
    For j = 1 To N + 1
      temp = A(p, j): A(p, j) = A(i, j): A(i, j) = temp
    Next j
    For i = p + 1 To N
      m(i) = Application.ImDiv(A(i, p), A(p, p))
    Next i
    For i = p + 1 To N
      For j = 1 To N + 1
        mA = Application.ImProduct(m(i), A(p, j))
        A(i, j) = Application.ImSub(A(i, j), mA)
      Next j
    Next i
  Next p
  For p = N To 2 Step -1                                      ' Step 3
    For i = 1 To p - 1
      If Application.ImAbs(A(p, p)) < tol Then A(p, p) = zeroC
      If Application.ImAbs(A(p, p)) = 0 Then
        fail = 1: Exit Sub
      End If
      m(i) = Application.ImDiv(A(i, p), A(p, p))
    Next i
    For i = 1 To p - 1
      For j = p To N + 1
        mA = Application.ImProduct(m(i), A(p, j))
        A(i, j) = Application.ImSub(A(i, j), mA)
      Next j
    Next i
  Next p
  For i = 1 To N                                              ' Step 4
    x(i) = Application.ImDiv(A(i, N + 1), A(i, i))
  Next i
End Sub  ' lineq
``` |

Example Call 5: Use Program **lineq** to solve:

$$\begin{bmatrix} 1+j1 & 2-j2 \\ 2+j2 & -1+j3 \end{bmatrix} \begin{bmatrix} x_1 \\ x_2 \end{bmatrix} = \begin{bmatrix} -7-j17 \\ -11+j21 \end{bmatrix}.$$

The following table shows the spreadsheet for Program **lineq**.

Example Call 5 Spreadsheet								
Input								Output	
2								
1	2	-7		1	-2	-17	-4	5
2	-1	-11		2	3	21	5	-4

The solution is: $\begin{bmatrix} x_1 \\ x_2 \end{bmatrix} = \begin{bmatrix} -4+j5 \\ 5-j4 \end{bmatrix}.$

Program inverse for Matrices with Complex Elements

For this program, the partitioned matrix is $[\ A\ \vdots\ I\ ]$, where A is the matrix to be inverted and *I* is the identity matrix. By successive row operations, this matrix is transformed to $[\ I\ \vdots\ A^{-1}\ ]$. These operations effectively do the following:

- Step 1: $[\ A\ \vdots\ I\ ]$
- Steps 2 and 3: $[\ A^{-1}A\ \vdots\ A^{-1}I\ ]$
- Step 4: $[\ I\ \vdots\ A^{-1}\ ]$

Example Call 6: Using row operations, invert the following matrix:

$$A = \begin{bmatrix} 0 & 1/2 & 1/3 \\ 1/2 & 1/3 & 1/4 \\ 1/3 & 1/4 & 1/5 \end{bmatrix}.$$

The following tables show the process.

Step 1 Form the partitioned matrix	0	1/2	1/3	1	0	0
	1/2	1/3	1/4	0	1	0
	1/3	1/4	1/5	0	0	1

Step 2.1 Because of the zero on the diagonal, interchange rows 1 and 2.	1/2	1/3	1/4	0	1	0
	0	1/2	1/3	1	0	0
	1/3	1/4	1/5	0	0	1

Step 2.2 On the matrix in Step 2.1, multiply row 1 by 2/3 and subtract it from row 3.	1/2	1/3	1/4	0	1	0
	0	1/2	1/3	1	0	0
	0	1/36	1/30	0	-2/3	1

Step 2.3 Multiply row 2 by 1/18 and subtract it from row 3.	1/2	1/3	1/4	0	1	0
	0	1/2	1/3	1	0	0
	0	0	1/67.5	-1/18	-2/3	1

Step 3.1 Multiply row 3 by 16.88 and subtract it from row 1.	1/2	1/3	0	0.938	12.25	-16.875
	0	1/2	1/3	1	0	0
	0	0	1/67.5	-1/18	-2/3	1

Step 3.2 Multiply row 3 by 22.5 and subtract it from row 2.	1/2	1/3	0	0.938	12.25	-16.875
	0	1/2	0	2.25	15	-22.5
	0	0	1/67.5	-1/18	-2/3	1

| Step 3.3
Multiply row 2 by 2/3 and subtract it from row 1.	1/2	0	0	-0.563	2.25	-1.875
	0	1/2	0	2.25	15	-22.5
	0	0	1/67.5	-1/18	-2/3	1

| Step 4
Divide row 1 by 1/2, row 2 by 1/2, and row 3 by 1/67.5.	1	0	0	-1.125	4.5	-3.75
	0	1	0	4.5	30	-45
	0	0	1	-3.75	-45	67.5

The inverse is: $A^{-1} = \begin{bmatrix} -1.125 & 4.5 & -3.75 \\ 4.5 & 30 & -45 \\ -3.75 & -45 & 67.5 \end{bmatrix}$.

The following table is a listing of Program **inverse**. In the program, A is the name of the partitioned matrix. Note that the program calls Program **mult** to check its answer.

The Call Statement

Program inverse to Invert a Matrix with Complex Elements

Code page 1

```
Sub inverse_main()
  Dim A(20, 50), ainv(20, 20), aainv(20, 20)
  Dim Ain(20, 20), ainR(20, 20), ainI(20, 20)
  Dim aR(20, 50), aI(20, 50)
  N = Cells(1, 1)
  For i = 1 To N
    For j = 1 To N
      ainR(i, j) = Cells(i + 1, j)
      ainI(i, j) = Cells(i + 1, j + N + 1)
      Ain(i, j) = Application.Complex(ainR(i, j), ainI(i, j))
    Next j
  Next i
  For i = 1 To N
    For j = 1 To N + N
      aR(i, j) = 0
      aI(i, j) = 0
    Next j
  Next i
  For i = 1 To N
    For j = 1 To N
      aR(i, j) = ainR(i, j): aI(i, j) = ainI(i, j)
    Next j
  Next i
  For i = 1 To N
    aR(i, i + N) = 1
  Next i
  For i = 1 To N
    For j = 1 To N + N
      A(i, j) = Application.Complex(aR(i, j), aI(i, j))       ' Step 1
    Next j
  Next i
  Call inverse(A, ainv, N, fail)
  If fail = 1 Then
    Cells(N + 3, 1) = "no solution": End
  End If
  For i = 1 To N
    For j = 1 To N
      ainvR = Application.ImReal(ainv(i, j))
      ainvI = Application.Imaginary(ainv(i, j))
      Cells(N + 3 + i, j) = Application.Round(ainvR, 3)
      Cells(N + 3 + i, j + N + 1) = Application.Round(ainvI, 3)
    Next j
  Next i
  Call mult(Ain, ainv, aainv, N, N, N)
  For i = 1 To N
    For j = 1 To N
      aainvR = Application.ImReal(aainv(i, j))
      aainvI = Application.Imaginary(aainv(i, j))
      Cells(2 * N + 5 + i, j) = aainvR
      Cells(2 * N + 5 + i, j + N + 1) = aainvI
    Next j
  Next i
  Cells(N + 3, 1) = "ainv(real)"
  Cells(N + 3, N + 2) = "ainv(imag)"
  Cells(2 * N + 5, 1) = "A*Ainv(real)"
  Cells(2 * N + 5, N + 2) = "A*Ainv(imag)"
End Sub  ' inverse_main
```

Code page 2

```
Sub inverse(A, ainv, N, fail)
  Dim m(50)
  tol = 0.000001: zeroC = Application.Complex(0, 0)
  fail = 0
  For p = 1 To N - 1 Step 1                                    ' Step 2
    For i = p To N
      If Application.ImAbs(A(i, p)) < tol Then A(i, p) = zeroC
      If Application.ImAbs(A(i, p)) <> 0 Then Exit For
      If i = N Then
        fail = 1: Exit Sub
      End If
    Next i
    For j = 1 To N + N
      temp = A(p, j): A(p, j) = A(i, j): A(i, j) = temp
    Next j
    For i = p + 1 To N
      m(i) = Application.ImDiv(A(i, p), A(p, p))
    Next i
    For i = p + 1 To N
      For j = 1 To N + N
        mA = Application.ImProduct(m(i), A(p, j))
        A(i, j) = Application.ImSub(A(i, j), mA)
      Next j
    Next i
  Next p
  For p = N To 2 Step -1                                       ' Step 3
    For i = 1 To p - 1
      If Application.ImAbs(A(p, p)) < tol Then A(p, p) = zeroC
      If Application.ImAbs(A(p, p)) = 0 Then
        fail = 1: Exit Sub
      End If
      m(i) = Application.ImDiv(A(i, p), A(p, p))
    Next i
    For i = 1 To p - 1
      For j = p To N + N
        mA = Application.ImProduct(m(i), A(p, j))
        A(i, j) = Application.ImSub(A(i, j), mA)
      Next j
    Next i
  Next p
  For i = 1 To N                                               ' Step 4
    For j = N + 1 To N + N
      A(i, j) = Application.ImDiv(A(i, j), A(i, i))
    Next j
    A(i, i) = Application.Complex(1, 0)
  Next i
  For i = 1 To N
    For j = N + 1 To N + N
      ainvR = Application.ImReal(A(i, j))
      ainvI = Application.Imaginary(A(i, j))
      ainv(i, j - N) = Application.Complex(ainvR, ainvI)
    Next j
  Next i
End Sub  ' inverse
Sub mult
  See this chapter.
End Sub
```

Example Call 7: Use Program **inverse** to invert the following matrix:

$$A = \begin{bmatrix} 1+j1 & 2-j2 \\ 2+j2 & -1+j3 \end{bmatrix}.$$

The following table shows the spreadsheet for Program **inverse**.

Example Call 7 Spreadsheet							
Input				Output			
2							
1	2	1	-2	0.122	0.189	-0.23	-0.135
2	-1	2	3	0.135	-0.068	0.189	-0.095

The inverse is:
$$A^{-1} = \begin{bmatrix} 0.122 - j0.23 & 0.189 - j0.135 \\ 0.135 + j0.189 & -0.068 - j0.095 \end{bmatrix}.$$

Chapter 11: The GoSub Statement and Runge-Kutta Numerical Integration

This chapter exhibits another way a program can be modularized. The following table shows a program that demonstrates this.

GoSub Demonstration Program	
Code	Comments

Sub main()
 	
    a = 10
    GoSub one
    GoSub two
    End
 	
one:
    Cells(1, 1) = a
    Return
 	
two:
    b = 20
    Cells(2, 1) = b
    Return
 	
End Sub
	• The statement, GoSub one, sends the program to the statement labeled *one*. There, a is printed. Then the Return statement sends the program back to the statement after GoSub one.
	
	• GoSub two sends the program to the statement labeled *two*. There, b is set to 20 and is printed. Then the Return statement sends the program back to the statement after GoSub two.
	
	• End terminates the program. This prevents looping.

The Syntax for a Statement Label: It must be some unique name followed by a *colon*. It must be the first entry on a line. (VBA will reposition it to column 1.) There's no limit on the number of labels.	
Note: The GoSub statement may also be used in a subprogram or a function.	

The following example fits into this format.

Example GoSub: Using Runge-Kutta integration, determine the response of the following system of equations:

$$\dot{Y}_1 = Y_2$$
$$\dot{Y}_2 = K_1*(Y_3 - Y_1) - K_2 Y_2$$
$$\dot{Y}_3 = K_3*(U_{in} - Y_3)$$
$$Y_{out} = Y_1 \quad \text{and} \quad \dot{Y}_{out} = Y_2$$

The System Equations

U_{in} is unit step function, and K_1, K_2, and K_3 are constants. Plot Y_{out} and its derivative as a function of time.

The program will use the **fourth-order Runge-Kutta** numerical-integration method. This very popular method is well documented. It is based on approximating the solution using a fourth-order **Taylor** series. Mathews and Fink show the 11 nonlinear equations with 13 unknowns that **Runge-Kutta** solved to yield the four coefficients in their method.[9]

Besides the differential equations and the integrator equations, the program will need the following:

- The initial conditions on time and the integrator outputs (Y_1, Y_2, and Y_3)
- The integration step size
- The stop time
- The variables to be output

The program will be designed as follows. It will have one main part and three segments:

- The main part will be the statements directing execution and dealing with input/output.
- A segment that we will call **runge** for the integrator statements.
- Another segment called **system** for the equation statements.
- A third segment called **data** for the statements that specify all constants.

The following is a block diagram of the program.

---

[9] J. Mathews and K. Fink, *Numerical Methods Using Matlab* (Upper Saddle River, NJ: Prentice Hall, 2004).

The GoSub Statement

The GoSub data statement enters the constants.

⇓

Integrate the equations, and save the output variables.

For i = 1 To (the number of time increments between start and stop)
  Do
    GoSub system (compute the derivatives)
    GoSub runge (integrate the derivatives)
  Loop
  After each time increment, save the output.
Next i

⇓

Print out.

⇓

The End statement terminates the program.

---

data:
- system constants K1 = 2, K2 = 0.5, K3 = 10, U_{in} = 1
- system order n=3
- initial conditions time h=0, y(1)=0, y(2)=0, y(3) = 0
- integration data: time increment dt=.05, tstop=15, istop = int(tstop/dt), where int yields the nearest integer in the negative direction
- integrator constants runm, jint
- initial values on the output variables
  - time: timeX(1) = 0
  - yout: Xy1(1) = y(1) = 0
  - yout_rate: Xy2(1) = y(2) = 0

Return

---

system:

$$\dot{Y}_1 = Y_2$$

$$\dot{Y}_2 = K_1 * (Y_3 - Y_1) - K_2 Y_2$$

$$\dot{Y}_3 = K_3 * (U_{in} - Y_3)$$

$$Y_{out} = Y_1 \quad \text{and} \quad \dot{Y}_{out} = Y_2$$

Return

---

runge:
    The Runge-Kutta integrator equations compute the derivatives four times in each time increment.
Return

The GoSub Statement

The following is a listing of the program.

Program for Example GoSub	
| ```
Option Base 1
Sub GoSub_main()
  Dim timeX(2000), y(20), yo(20), yd(20), runk(20, 4)
  Dim xy1(2000), xy2(2000)
  GoSub data
  For i = 1 To (istop + 1)
    Do
      GoSub system
      GoSub runge
    Loop While jint <> 1
    timeX(i + 1) = timeh
    xy1(i + 1) = yout : xy2(i + 1) = yout_rate
  Next i
  For i = 1 To (istop + 1)
    Cells(i + 5, 1) = timeX(i)
    Cells(i + 5, 2) = xy1(i): Cells(i + 5, 3) = xy2(i)
  Next i
End
data:
  N = 3: y(1) = 0: y(2) = 0: y(3) = 0
    timeh = 0: dt = 0.05: tstop = 15: istop = Int(tstop / dt)
    yout = y(1): yout_rate = y(2)
    timeX(1) = timeh: xy1(1) = yout: xy2(1) = yout_rate
    runm = Array(1, 0.5, 0.5, 1): jint = 1
    K1 = 2: K2 = 0.5: K3 = 10: Uin = 1
Return
system:
  yd(1) = y(2)
  yd(2) = -K2 * y(2) + K1 * (y(3) - y(1))
  yd(3) = K3 * (Uin - y(3))
  yout = y(1)
  yout_rate = y(2)
Return
runge:
For j = 1 To N
  runk(j, jint) = dt * yd(j)
Next j
If jint = 1 Then
  tox = timeh
  For j = 1 To N
    yo(j) = y(j)
  Next j
End If
jint = jint + 1
If jint > 1 And jint < 5 Then
  timeh = tox + runm(jint) * dt
  For j = 1 To N
    y(j) = yo(j) + runm(jint) * runk(j, jint - 1)
  Next j
End If
If jint = 5 Then
  jint = 1
  For j = 1 To N
    inc = (runk(j, 1) + 2 * runk(j, 2) + 2 * runk(j, 3) + runk(j, 4)) / 6
    y(j) = yo(j) + inc
  Next j
End If
Return
End Sub ' GoSub_main
``` | • The GoSub data statement.<br><br><br><br>• Integrate and save the output variables.<br><br><br><br>• Printout.<br><br>• The End statement terminates the program.<br><br><br><br><br>• data: ... Return<br><br><br><br><br>• system: ... Return (*Note:* The equations in this segment may be nonlinear.)<br><br>• runge: ... Return<br><br><br>• The system derivatives are computed four times for each dt.<br><br>  • jint = 1 initialize the integrator.<br><br>  • jint = 2 compute the first estimate of the y vector.<br><br>  • jint = 3 compute the second estimate of the y vector.<br><br>  • jint = 4 compute the third estimate of the y vector.<br><br>  • jint = 5 compute the fourth estimate and then the weighted average of the estimates. Reset jint = 1 for the next dt.<br><br>• *Note:* The above data and system segments produced the results plotted on the next page. They can be replaced by other segments. See chapter 15 and appendix E. |
```

The following table shows a partial list of the data output to the spreadsheet.

Step Response for Example GoSub		
time, timeX(j)	yout, Xy1(j)	yout_rate, Xy2(j)
0	0	0
0.05	0.001	0.018
0.1	0.003	0.071
0.15	0.008	0.139
0.2	0.017	0.216
•	•	•
•	•	•
•	•	•
14.8	0.996	0.037
14.85	0.998	0.036
14.9	1.000	0.035
14.95	1.002	0.034
15	1.003	0.033

The following are the scatter charts of the responses.

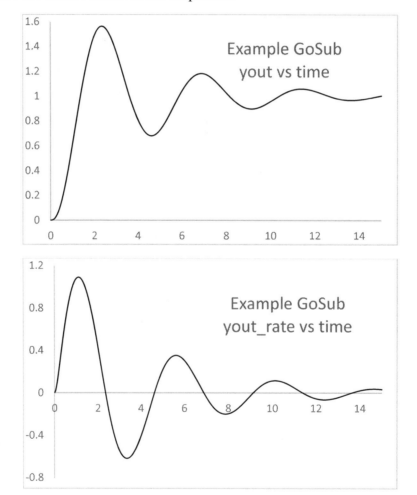

The above plot of Y_{out} also appears in chapter 1, **example 2**. Throughout this book, all plots that are referred to as being from **Runge-Kutta** integration were made using the program in **Example GoSub**. Of course, the program segments, data and system, were modified accordingly.

Chapter 12: The GoTo Statement with Sort and Interpolate

This chapter defines the GoTo statement. Simply stated, GoTo acts like GoSub without the Return. The following table contains a demonstration program.

GoTo Demonstration Program	
Code	Comments

Sub main()
    GoTo one
 	
two:
    Cells(1, 1) = a
    End
 	
one:
    a = 10
    GoTo two
End Sub
	• GoTo one sends the program to the statement labeled one. There, a is set to 10. Then the GoTo two statement sends the program to the statement labeled two.
	
	• At two, a is printed.
	
	• End terminates the program. This prevents looping.

The Syntax for a Statement Label It must be some unique name followed by a *colon*. It must be the first entry on a line. (VBA will reposition it to column 1.) There's no limit on the number of labels.	
Note: The GoTo statement may also be used in a subprogram or function.	

The following example uses the GoTo statement.

Example GoTo: Take a list of paired numbers, Xin(i) and Yin(i). Sort Xin(i) in ascending order.

Unsorted		Sorted	
Xin(i)	Yin(i)	Xin(i)	Yin(i)
9	3	-9	-3
-9	-3	-7	-1
7	1	-5	1
-7	-1	-3	3
5	1	-1	5
-5	1	1	5
3	3	3	3
-3	3	5	1
1	5	7	1
-1	5	9	3

Then by interpolating in this sorted list, find the **y** that corresponds to a value input for x. For this example, use x = - 4.

The following table lists the program for **Example GoTo**. The GoTo statement is used in the interpolation segment.

The GoTo Statement

Program for Example GoTo	
Code	Comments

Sub main()
  Dim Xin(100), Yin(100)
  n = Cells(1, 1): x = Cells(1, 6)
  For i = 1 To n
    Xin(i) = Cells(i, 2): Yin(i) = Cells(i, 3)
  Next i
  For L = n To 2 Step -1
    For i = 1 To L - 1
      If Xin(i) > Xin(L) Then
        temp = Xin(i)
        Xin(i) = Xin(L)
        Xin(L) = temp
        temp = Yin(i)
        Yin(i) = Yin(L)
        Yin(L) = temp
      End If
    Next i
  Next L
  For i = 1 To n
    Cells(i, 4) = Xin(i): Cells(i, 5) = Yin(i)
  Next i
  If (x < Xin(1)) Or (x > Xin(n)) Then
    Cells(1,7) = " x out of bounds": End
  End If
  For i = 2 To n
    If x <= Xin(i) Then
      m = i - 1
      GoTo st200
    End If
  Next i
st200:
  ydel = (Yin(m + 1) - Yin(m)) * (x - Xin(m)) / _
      (Xin(m + 1) - Xin(m))
  y = Yin(m) + ydel
  Cells(1, 7) = y
End Sub
	• n is the number of points and x *is the Input*.
	• Unsorted Pair.
	
	• Sort:
	
	  • Sort into ascending order.
	  • Outer loop runs from bottom up.
	  • Inner loop runs from top down.
	  • First pass puts largest number on the bottom.
	  • Second pass puts second-largest number second last.
	  • This pattern continues.
	
	• Print the sorted pair.
	
	• Check the bounds.
	
	• Interpolate:
	
	Beginning at the smallest number,
	when x is between Xin(m) and Xin(m+1),
	the GoTo st200 statement transfers
	execution to the statement labeled st200.
	
	• Linear interpolation.
	
	• y is the Output that corresponds to *Input x*.

Syntax: A space followed by an underscore (_) allows a line to continue on the next line.	

The following is the **Scatter** chart of the sorted input list and the result.

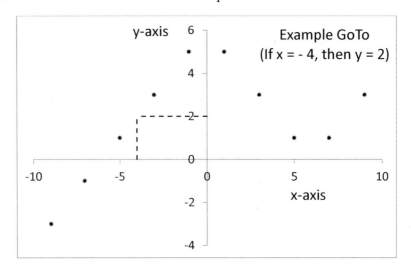

Chapter 13: Debugging

Debugging is fun. Point-and-click menus and windows make it easy. Execution can be stopped in several ways and at any line of code. Values of variables can be inspected in several ways. Execution can then be continued until the next stopping point. A simple program shows all of this. (This chapter may be skipped over until needed.)

The following table shows the listing and description of the program that will be analyzed by debugging. The program merely passes a variable around a loop between a main program and two subprograms. Type this program into a code window.

The Program Used to Demonstrate Debugging	
Code	Description

`Sub main()`
`    For w = 1 To 5`
`        Call one(w, zDum)`
`        Cells(w, 1) = zDum`
`    Next w`
`End Sub`
 	
`Sub one(wDum, z)`
`    x = wDum`
`    Call two(x, yDum)`
`    z = yDum`
`End Sub`
 	
`Sub two(xDum, y)`
`    y = xDum`
`End Sub`
	• In Sub main, the For statement starts with w = 1.
	• Call Sub one with w in the argument list.
	• Sub one
	  • wDum is equivalenced to w in the argument list.
	  • x = wDum.
	  • x is passed to Sub two.
	• Sub two
	  • xDum is equivalenced to x.
	  • y = xDum.
	  • y is passed back to Sub one.
	• Sub one
	  • yDum is equivalenced to y.
	  • z = yDum.
	  • z is passed back to Sub main
	• Sub main
	  • zDum is equivalenced to z.
	  • zDum is printed out.
	• The Next statement then sets w = 2, and execution continues on the second pass.

The first step in debugging is to open the **Watches** window.

> Click **Editor**, and then in the **View** menu, click **Watch Window**.

Watches will show the values of selected variables as the program executes. For this demonstration, all the variables in the above program will be put into **Watches**. Following are the instructions on how to do this.

- Open the code window that contains the above program.
- Put the cursor next to a variable (e.g., w).
- Then in the **Debug** menu, click **Add Watch**. The **Add Watch** window pops up, and the variable w is already entered. (Options in the **Add Watch** window are discussed later.)
- Click OK. The **Add Watch** window closes. The variable w now appears in **Watches** and is out of context since the program is not yet executing.
- Put the cursor next to another variable (e.g., zDum). Repeat the process until **Watches** looks like the following table.

The Watches Window before Execution			
Expression	Value	Type	Context
w	Out of Context	Variant/Empty	ThisWorkbook.main
zDum	Out of Context	Variant/Empty	ThisWorkbook.main
z	Out of Context	Variant/Empty	ThisWorkbook.one
wDum	Out of Context	Variant/Empty	ThisWorkbook.one
x	Out of Context	Variant/Empty	ThisWorkbook.one
yDum	Out of Context	Variant/Empty	ThisWorkbook.one
xDum	Out of Context	Variant/Empty	ThisWorkbook.two
y	Out of Context	Variant/Empty	ThisWorkbook.two
Note: The Watches window is not saved when the program file is closed.			

Variables can be added to the **Watches** window at any time during debug.

Now to begin debugging, open the code window, and put the cursor anywhere in **Sub main**. Select the **Debug** menu. The focus is now on four commands: **Step Into**, **Step Over**, **Step Out**, and **Run To Cursor**. We will start with the **Step Into** command.

- **Step Into**: In the **Debug** menu, click **Step Into**. This **Step Into** command puts the arrow on line 1 of the code. At the next **Step Into** command, line 1 will execute, and the cursor will point to line 2. The following two tables show each successive **Step Into** command and the values of the variables after each step.

This table shows the code line after each **Step Into** command.

The Code Window	
Step Into *Command Numbers*	Code Lines
1	Sub main()
2	For w = 1 To 5
3 and 14	Call one(w, zDum)
12	Cells(w, 1) = zDum
13	Next w
	End Sub
4	Sub one(wDum, z)
5	x = wDum
6	Call two(x, yDum)
10	z = yDum
11	End Sub
7	Sub two(xDum, y)
8	y = xDum
9	End Sub

This table lists the value of the variables after each **Step Into** command.

		The Watches Window														
		The Value *after* Each Step Into Command														
	Start	1	2	3	4	5	6	7	8	9	10	11	12	13	14
w	O	E	E	1	1	1	1	1	1	1	1	1	1	1	2
zDum	O	E	E	E	E	E	E	E	E	E	E	1	1	1	1
z	O	O	O	O	E	E	E	E	E	E	E	1	O	O	O
wDum	O	O	O	O	1	1	1	1	1	1	1	1	O	O	O
x	O	O	O	O	E	E	1	1	1	1	1	1	O	O	O
yDum	O	O	O	O	E	E	E	E	E	1	1	1	O	O	O
xDum	O	O	O	O	O	O	O	1	1	1	O	O	O	O	O
y	O	O	O	O	O	O	O	E	E	1	O	O	O	O	O
Notes: • O legend for Out of context, i.e., in a part of the program that is not active • E legend for Empty																

The tables show that after the third **Step Into** command, line 2 has executed, and w = 1. Continue analyzing the above table, or go to the **Run** menu and **Reset**.

Debugging

The Other Options in the Debug Menu

• **Step Over**: When the Debug arrow is pointing to a Call statement (or a Function statement), the Step Over command causes the debugger to execute the called program and stop on the statement after the Call.

• **Step Out**: When the Debug arrow is pointing to any statement in a subprogram or function, the Step Out command causes the debugger to execute the rest of the subprogram and stop on the statement after the Call. Note that when the Debug arrow is pointing to any statement in the main program, the Step Out command causes the rest of the program to be completed and stop.

• **Run To Cursor**: Put the cursor at the beginning or end of any statement. The Run To Cursor command causes the debugger to execute all statements up to the cursor and stop at the cursor.

Viewing Variables with the Cursor

This is another way to see the values of variables. Suppose the Debug arrow is pointing to the following statement: Call one(w, zDum). Hold the mouse pointer over the variable w. After a slight delay, a pop-up will show the value of w. In fact, holding the mouse pointer over any variable for a moment will produce a pop-up showing the value, if it's available.

Debugging by Run-to-Breakpoint

Another way of debugging is to set breakpoints. The following table describes how to set breakpoints.

Code Window Illustrating Breakpoints	
Code	Description

Sub main()
  For w = 1 To 5
    Call one(w, zDum)
    Cells(w, 1) = zDum
  Next w
End Sub
Sub one(wDum, z)
•   x = wDum
  Call two(x, yDum)
  z = yDum
End Sub
Sub two(xDum, y)
•   y = xDum
End Sub
	• A breakpoint is set by clicking in the gray area in front of a line of code.
	
	• Two breakpoints (bullets) are shown.
	
	• Like a toggle switch, another click will turn it off.
	
	• The program need not be in the debug mode when setting a breakpoint.

Debugging

With the breakpoints as shown, Run the program. Execution will stop at the statement x = wDum. Values of the variables can be inspected either in Watches or with the mouse pointer. Then Run Menu → Continue causes execution to stop at the next breakpoint. A debugging session can include using the options from the Debug menu together with run-to-breakpoint. The menus always show the options that are available. When debugging is finished, select Run Menu → Reset

Options in the Add Watch Window

Besides variables, the Watches window can include arithmetic, comparative, and logical expressions (see chapter 9). The Add Watch window allows one of the following options to be selected for each expression:

- Watch Expression (the option used so far in this chapter)
- Break When Value Is True
- Break When Value Changes

The following are examples of expressions that can be put into the Watches window. Never mind if they don't make sense.

w	a variable
x + wDum - y * z	an arithmetic expression
y > 3	a comparison expression
w > 3 And zDum > 3	a logical expression

The Immediate Window

This window can display program output via special VBA debug statements that are placed in the code. This window will not be discussed in this book.

Final Note

Back to the oxymoron that *debugging is fun*. When the code malfunctions, VBA automatically stops in the debug mode, points an arrow at the difficult line of code, and allows values to be inspected by hovering the cursor above any variable. And, of course, the debug mode can be used as a learning aid even when there is no malfunction.

Chapter 14: VBA Programs to Compute Eigenvalues and Eigenvectors

All of the syntax needed to write VBA programs in *Excel* has now been presented. The two main programs in this book can now be shown.

1. A Program to Compute All of the Eigenvalues of a General Real Matrix

A general real matrix has complex eigenvalues. The procedure that has been programmed is shown in the flowchart on the next page. The following is a higher-level version of that flowchart.

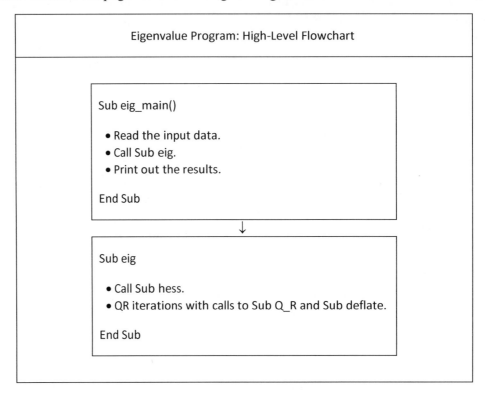

This chapter discusses the listing of **Sub eig_main** and its input and output. The listings of the rest of the subprograms are in appendix D.3.

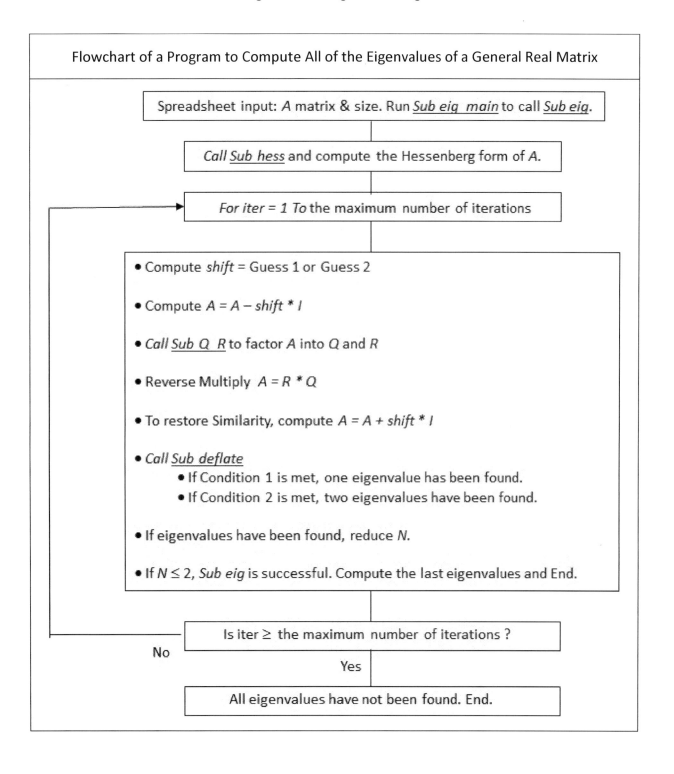

The tables on the next page show the use of program **Sub eig_main**. The top table is an annotated listing of the program. The bottom table is its spreadsheet.

The input A matrix is from example 2 in chapter 1.

$$A = \begin{bmatrix} 0 & 1 & 0 \\ -2 & -0.5 & 2 \\ 0 & 0 & -10 \end{bmatrix}$$

The output includes the eigenvalues: $\lambda_1 = -10$, $\lambda_2 = -0.25 + j1.39$, and $\lambda_3 = -0.25 - j1.39$. Four additional variables are computed and printed to the spreadsheet. These are the following:

- **eigsum**, which is the sum of the eigenvalues
- **trace**, which is the trace of the A matrix
- **prodR**, which is the product of the eigenvalues
- **deter**, which is the determinant of the A matrix

Golub and Van Loan prove the following two properties of eigenvalues:[10]

- The trace of a matrix equals the sum of its eigenvalues.
- The determinant of a matrix equals the product of its eigenvalues.

Therefore, if the eigenvalues are correct, **eigsum** will equal **trace**, and **prodR** will equal **deter**.

---

[10] G. Golub and C. Van Loan, *Matrix Computations* (Baltimore, MD: Johns Hopkins University Press, 1989).

Listing of **Sub eig_main** and the spreadsheet of its input and output.

Listing of *Sub eig_main*	
Code	Comments
| ```
Option Base 1
Sub eig_main()
Dim A(), WR(30), WI(30)
N = Cells(1, 1): ReDim A(N, N)
For i = 1 To N
  For j = 1 To N
    A(i, j) = Cells(i + 1, j)
  Next j
Next i
Call eig(N, A, WR, WI)
For i = 1 To N
  Cells(1 + i, N + 2) = WR(i): Cells(1 + i, N + 3) = WI(i)
Next i
eigsum = 0: trace = 0: prod = Application.Complex(1, 0)
psign = 1
For i = 1 To N
  eigsum = eigsum + WR(i)
  trace = trace + A(i, i)
  Root = Application.Complex(WR(i), WI(i))
  prod = Application.ImProduct(prod, Root)
  prodR = Application.ImAbs(prod)
  psign = psign * WR(i)
Next i
  If psign < 0 Then prodR = -prodR
  deter = Application.MDeterm(A)
Cells(N + 4, 1) = Application.Round(eigsum, 3)
Cells(N + 4, 2) = Application.Round(trace, 3)
Cells(N + 4, 3) = Application.Round(prodR, 3)
Cells(N + 4, 4) = Application.Round(deter, 3)
Cells(1, N + 2) = "   WR(i)": Cells(1, N + 3) = "   WI(i)"
Cells(N + 3, 1) = "   eigsum": Cells(N + 3, 2) = "   trace"
Cells(N + 3, 3) = "   prod": Cells(N + 3, 4) = "   deter"
End Sub        ' eig_main
``` | • Read N, the size of the A matrix.<br><br><br><br>• Read the A matrix.<br><br>• Call Sub eig<br><br>• Print out the eigenvalues. $\lambda_i = WR_i + j\,WI_i$<br><br>• Compute and print out<br><br>    • eigsum<br>    • trace<br>    • prodR<br>    • deter |
| <u>Note</u>: Sub eig_main, Sub eig, Sub hess, Sub deflate, and Sub Q_R must be in the same code window. The listings of these programs are in Appendix D.3. ||

The Spreadsheet of *Sub eig_main* for Example 2 in Chapter 1				
3			WR(i)	WI(i)
0	1	0	-10	0
-2	-0.5	2	-0.25	1.391941
0	0	-10	-0.25	-1.39194
eigsum	trace	prod	deter	
-10.5	-10.5	-20	-20	

• The eigenvalues are: $\lambda_1 = -10$, $\lambda_2 = -0.25 + j\,1.39$, $\lambda_3 = -0.25 - j\,1.39$
• For checkout: **eigsum** = **trace** and **prod** = **deter**.
```

2. A Program to Compute All of the Eigenvectors of a General Real Matrix

A general real matrix has complex eigenvectors. The procedure that has been programmed is shown in the flowchart on the next page. The following is a higher-level version of that flowchart.

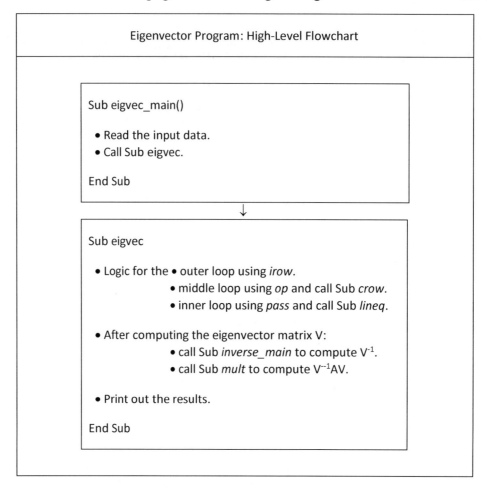

This section discusses the listing of **Sub eigvec_main** and its input and output. The listings of the rest of the subprograms are in appendix D.4.

The Eigenvalue and Eigenvector Programs

Flowchart of a Program to Compute All of the Eigenvectors of a General Real Matrix

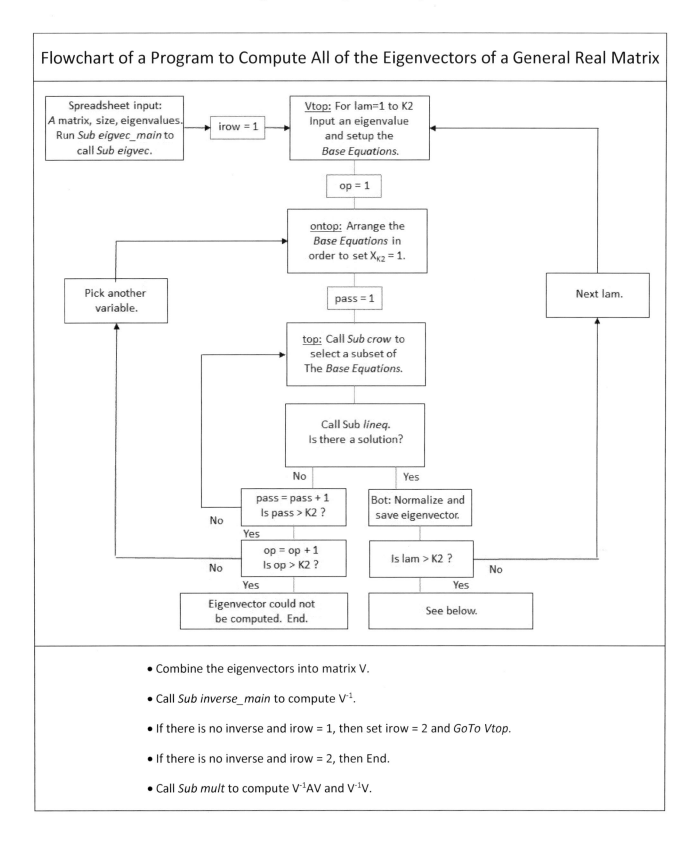

- Combine the eigenvectors into matrix V.
- Call *Sub inverse_main* to compute V^{-1}.
- If there is no inverse and irow = 1, then set irow = 2 and *GoTo Vtop*.
- If there is no inverse and irow = 2, then End.
- Call *Sub mult* to compute $V^{-1}AV$ and $V^{-1}V$.

The tables on the next page show how to use program **Sub eigvec_main**. The top table is a listing of the program. The bottom table shows its spreadsheet.

The A matrix and its eigenvalues are from example 2 in chapter 1.

$$A = \begin{bmatrix} 0 & 1 & 0 \\ -2 & -0.5 & 2 \\ 0 & 0 & -10 \end{bmatrix} \qquad \lambda_1 = -10, \; \lambda_2 = -0.25 + j1.39, \; \lambda_3 = -0.25 - j1.39.$$

The output eigenvector matrix is:

$$V = \begin{bmatrix} 0.0206 & -0.125 - j0.696 & -0.125 + j0.696 \\ -0.2026 & 1 & 1 \\ 1 & 0 & 0 \end{bmatrix}.$$

This agrees with the results in example 2 in chapter 1.

Also output is:

$$V^{-1} = \begin{bmatrix} 0 & 0 & 1 \\ j0.7184 & 0.5 + j0.0898 & 0.1031 + j0.0037 \\ -j0.7184 & 0.5 - j0.0898 & 0.1031 - j0.0037 \end{bmatrix}.$$

The **lamda** matrix is also output:

$$L = V^{-1}AV = \begin{bmatrix} -10 & 0 & 0 \\ 0 & -0.25 + j1.39 & 0 \\ 0 & 0 & -0.25 - j1.39 \end{bmatrix}.$$

Finally, the ouput includes:

$$V^{-1}V = \begin{bmatrix} 1 & 0 & 0 \\ 0 & 1 & 0 \\ 0 & 0 & 1 \end{bmatrix}.$$

Important Note: Owing to the output format that I selected, the zeros in the output matrices are actually numbers that are less than 0.00005.

The Eigenvalue and Eigenvector Programs

Listing of **Sub eigvec_main** and the spreadsheet of its input and output.

Listing of *Sub eigvec_main*	
Code	Comments

Option Base 1
 Sub eigvec_main()
   Dim arealR(20, 20), areal(20, 20), lamda(20)
   K2 = Cells(1, 1)
   arealI = Application.Complex(0, 0)
   For i = 1 To K2
     For j = 1 To K2
       arealR(i, j) = Cells(i + 1, j)
       areal(i, j) = Application.Complex(arealR(i, j), arealI)
     Next j
   Next i
   For i = 1 To K2
     lamdaR = Cells(i + 1, K2 + 2)
     lamdaI = Cells(i + 1, K2 + 3)
     lamda(i) = Application.Complex(lamdaR, lamdaI)
   Next i
   Call eigvec(K2, areal, lamda)
 End Sub ' eigvec_main
	• Read K2, the size of the A matrix.
	
	 • Read the A matrix and convert it to a complex matrix.
	
	 • Read the eigenvalue matrix and convert it to a complex matrix.
	
	 • Call Sub eigvec.

Note: Sub eigvec_main, Sub eigvec, Sub crow, Sub lineq, Sub inverse_main, Sub inverse, Sub mult, Sub coeff_step, Sub the_labels, Sub noinverse must be in the same code window. The listings of these programs are in Appendix D.4.

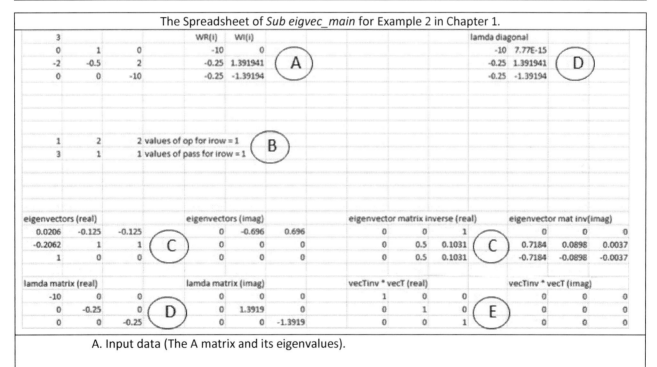

The Spreadsheet of *Sub eigvec_main* for Example 2 in Chapter 1.

A. Input data (The A matrix and its eigenvalues).

B. The flags (irow, op, and pass) when the eigenvector was computed for each eigenvalue.

C. The eigenvector matrix V and its inverse V^{-1}.

D. The matrix $L = V^{-1}AV$ and its diagonal, showing the accuracy of the eigenvectors.

E. The matrix $V^{-1}V$ for additional checkout.

See **example 16.3** for the printout when V cannot be inverted.

The Step Response of a Linear System Using Eigenvectors

This application of eigenvectors is important enough to be included in the program. If the B and C matrices are input as shown, the necessary coefficients are output.

The Spreadsheet of Sub eigvec_main for Example 2 in Chapter 1, with the B and C Matrices

3			WR(i)	WI(i)					lamda diagonal		
0	1	0	-10	0		0		1	-10	7.77E-15	
-2	-0.5	2	-0.25	1.391941	A	0		0	-0.25	1.391941	D
0	0	-10	-0.25	-1.39194		10		0	-0.25	-1.39194	

1	2	2 values of op for irow = 1	
3	1	1 values of pass for irow = 1	B

eigenvectors (real)			eigenvectors (imag)			eigenvector matrix inverse (real)			eigenvector mat inv(imag)		
0.0206	-0.125	-0.125	0	-0.696	0.696	0	0	1	0	0	0
-0.2062	1	1	0	0	0	0	0.5	0.1031	0.7184	0.0898	0.0037
1	0	0	0	0	0	0	0.5	0.1031	-0.7184	-0.0898	-0.0037

lamda matrix (real)			lamda matrix (imag)			vecTinv * vecT (real)			vecTinv * vecT (imag)		
-10	0	0	0	0	0	1	0	0	0	0	0
0	-0.25	0	0	1.3919	0	0	1	0	0	0	0
0	0	-0.25	0	0	-1.3919	0	0	1	0	0	0

Ac	Bc	WR	WI	
-0.0206	0	-10	0	
-0.4897	0.162	-0.25	1.3919	F
-0.4897	-0.162	-0.25	-1.3919	

F. When the B and C matrices are input, the step-response coefficients (A_C, B_C) are computed for the following equation (Appendix C, Equation 1):

$$Y = \sum_{i=1}^{3} \{e^{WR_i t}[Ac_i \cos(WI_i t) - Bc_i \sin(WI_i t)] - Ac_i\}$$

The step response for example 2 in chapter 1:

$$Y = e^{-10t}[-0.0206\cos(0) - 0\sin(0)] + 0.0206$$
$$+ e^{-0.25t}[-0.4897\cos(1.39t) - 0.162\sin(1.39t)] + 0.4897$$
$$+ e^{-0.25t}[-0.4897\cos(-1.39t) + 0.162\sin(-1.39t)] + 0.4897$$

$$Y = -0.0206 e^{-10t} + 2e^{-0.25t}[-0.4897\cos(1.39t) - 0.162\sin(1.39t)] + 1.0$$

For accuracy, this checks with the Runge-Kutta program in chapter 11.

Even if the step response is not the objective, it's a good way to check out the eigenvectors.

Chapter 15: Applications

This chapter shows results when the eigenvalue and eigenvector programs are applied to very coupled A matrices of actual systems. These systems are the following:

- Case 1: Yaw-roll airplane motion
- Case 2: Balancing a double inverted pendulum on a cart
- Case 3: Attitude control of a large, flexible launch vehicle
- Case 4: Mode shapes of a free vibrating system

The accuracy of the results is shown by the following internal cross-checks:

- Program Sub eig_main computes eigsum, trace, prodR, and deter.
- Program Sub eigvec_main computes lamda and $V^{-1}V$.

In addition, Runge-Kutta numerical integration is used to check the responses produced by each case.

The first three cases are applications that use the eigenvector matrix as a *matrix of rows*. That is, the eigenvectors show how all the eigenvalues combine to affect the response.

The last case uses the eigenvector matrix as a *matrix of columns*. That is, each eigenvector shows how the response of the system variables compares with each other at each eigenvalue. In this last case, the eigenvectors are called mode shapes.

Even if the eigenvalues and eigenvectors are intended for other applications, the way they are used in this chapter provides simple checks on their accuracy.

Applications

Case 1: Yaw-Roll Airplane Motion

The system equations for this case come from the standard math model of an airplane's yawing and rolling motions. These are called the lateral-directional modes of motion, and Stevens and Lewis provide a good reference.[11] The problem statement is to compute the eigenvalues and eigenvectors at the given flight condition and to use these to compute the response to a step roll-angle command of five degrees.

The system equations are in the following table, where

- ϕ (Fe) is the roll angle;
- p is the roll rate;
- r is the yaw rate;
- β is the sideslip angle;
- δ_r (Dr) is the rudder angle; and
- δ_a (Da) is the aileron angle.

The System Equations	
Motion Equations	Autopilot Equations
$\dot{\beta} = -r + \dfrac{g}{V}\cos(\theta_o)\phi + Y'_\beta \beta + Y'_{\delta r}\delta_r$ $\dot{r} = N'_p p + N'_r r + N'_\beta \beta + N'_{\delta r}\delta_r + N'_{\delta a}\delta_a$ $\dot{p} = L'_p p + L'_r r + L'_\beta \beta + L'_{\delta r}\delta_r + L'_{\delta a}\delta_a$	$\delta_a = k_\phi(\phi_c - \phi) - k_p p$ $\delta_r = k_r r$

The A and B matrices at the given flight condition are:

$$\begin{bmatrix} \dot{\beta} \\ \dot{r} \\ \dot{p} \\ \dot{p} \end{bmatrix} = \begin{bmatrix} -0.1276 & -0.9927 & 0.0731 & 0 \\ 1.6402 & -0.3384 & 0.0171 & -0.0405 \\ 0 & 0 & 0 & 1 \\ -2.0922 & 0.4904 & -2.2024 & -4.2629 \end{bmatrix} \begin{bmatrix} \beta \\ r \\ \phi \\ p \end{bmatrix} + \begin{bmatrix} 0 \\ -0.0171 \\ 0 \\ 2.2024 \end{bmatrix} \phi_c$$

---

[11] B. Stevens and F. Lewis, *Aircraft Control and Simulation* (New York: John Wiley and Sons, 1992).

Applications

The following table is the spreadsheet when the A matrix is input to **Program eig_main** to compute the eigenvalues.

Spreadsheet for Program Sub eig_main						
4 lat-dir	sheet 47				WR(i)	WI(i)
-0.12758	-0.99273	0.073123	0		-0.56994	0
1.640229	-0.33841	0.017086	-0.04046		-0.24123	1.295308
0	0	0	1		-0.24123	-1.29531
-2.09224	0.490374	-2.20235	-4.26289		-3.67647	0
eigsum	trace	prod	deter			
-4.729	-4.729	3.638	3.638			

The dutch-roll mode is given by $\lambda = -0.24 \pm j1.3$, and the combined roll-spiral-autopilot modes are given by $\lambda = -0.57$ and $\lambda = -3.68$.

The following table is the spreadsheet when the A matrix and the eigenvalues are input to **Program eigvec_main** to compute the eigenvectors.

Spreadsheet for Program Sub eigvec_main for φ Response															
4 lat-dir	sheet 40		Fe/Fec	WR(i)	WI(i)		" B "		" C "		lamda diagonal				
-0.12758	-0.99273	0.073123	0	-0.56994	0		0		0		-0.56994	7.07E-13			
1.640229	-0.33841	0.017086	-0.04046	-0.24123	1.295308		-0.017		0		-0.24123	1.295308			
0	0	0	1	-0.24123	-1.29531		0		5		-0.24123	-1.29531			
-2.09224	0.490374	-2.20235	-4.26289	-3.67647	0		2.2024		0		-3.67647	-2.4E-12			
1	1	1	1	values of op for irow = 1											
1	1	1	1	values of pass for irow = 1											
eigenvectors (real)				eigenvectors (imag)				eigenvector matrix inverse (real)				eigenvector mat inv(imag)			
0.0328	-0.7235	-0.7235	0.0082	0	0.2979	-0.2979	0	-0.2323	-0.3353	-1.1656	-0.312	0	0	0	0
-0.059	0.3012	0.3012	0.0095	0	0.9536	-0.9536	0	-0.6166	0.1806	-0.0343	-0.0059	-0.1842	-0.4568	0.0287	0.0136
-1	-0.0621	-0.0621	-0.272	0	-0.3336	0.3336	0	-0.6166	0.1806	-0.0343	-0.0059	0.1842	0.4568	-0.0287	-0.0136
0.5699	0.4472	0.4472	1	0	0	0	0	0.6838	0.0296	0.695	1.1831	0	0	0	0
lamda matrix (real)				lamda matrix (imag)				vecTinv * vecT (real)				vecTinv * vecT (imag)			
-0.5699	0	0	0	0	0	0	0	1	0	0	0	0	0	0	0
0	-0.2412	0	0	0	1.2953	0	0	0	1	0	0	0	0	0	0
0	0	-0.2412	0	0	0	-1.2953	0	0	0	1	0	0	0	0	0
0	0	0	-3.6765	0	0	0	0	0	0	0	1	0	0	0	0
Ac	Bc	WR	WI												
-5.9776	0	-0.5699	0												
0.0019	-0.0529	-0.2412	1.2953												
0.0019	0.0529	-0.2412	-1.2953												
0.9637	0	-3.6765	0												

Because the L (lamda) matrix is diagonal and shows the eigenvalues, the eigenvectors are accurate. Note that the zeros in L are actually numbers less than 0.00005. The eigenvectors for the dutch-roll mode have complex values, which is correct.

To compute the roll-angle response to a roll-angle step command, the B and C matrices are input as shown. The size of the command (5) is included in the C matrix. The program outputs the A_C and B_C coefficients. These are input to the following Y equation:

$$Y_\phi = -5.978e^{-0.57t} + 5.978$$
$$+ 2\{e^{-0.24t}[0.0019\cos(1.3t) + 0.053\sin(1.3t)] - 0.0019\}$$
$$+ 0.964e^{-3.68t} - 0.964$$

The contribution to the response, owing to each eigenvalue, is clearly shown by the eigenvectors. The following figure is a plot of Y_ϕ.

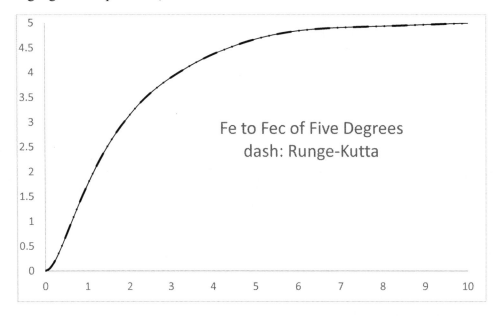

The plot also shows the response when the system equations are integrated using the Runge-Kutta method. This method is in the VBA program in chapter 11, and the data for doing this are given in appendix E. The comparison shows that Y_ϕ is correct.

The aileron response will not be shown, but it was also accurate.

To compute the rudder response to the roll-angle command, the C matrix is changed to that shown in the following table. The size of the command is included in the C matrix. (*Note:* the autopilot gain $k_r = 0.2$.) The eigenvectors have not changed because the A matrix has not changed.

Spreadsheet for Program Sub eigvec_main for D_r Response													
4 lat-dir		sheet 42		Dr/Fec	WR(i)	WI(i)	"B"	"C"		lamda diagonal			
-0.12758	-0.99273	0.073123	0		-0.56994	0	0	0		-0.56994	7.07E-13		
1.640229	-0.33841	0.017086	-0.04046		-0.24123	1.295308	-0.017	1		-0.24123	1.295308		
0	0	0	1		-0.24123	-1.29531	0	0		-0.24123	-1.29531		
-2.09224	0.490374	-2.20235	-4.26289		-3.67647	0	2.2024	0		-3.67647	-2.4E-12		
1	1	1	1 values of op for irow = 1										
1	1	1	1 values of pass for irow = 1										
eigenvectors (real)				eigenvectors (imag)				eigenvector matrix inverse (real)				eigenvector mat inv(imag)			
0.0328	-0.7235	-0.7235	0.0082	0	0.2979	-0.2979	0	-0.2323	-0.3353	-1.1656	-0.312	0	0	0	0
-0.059	0.3012	0.3012	0.0095	0	0.9536	-0.9536	0	-0.6166	0.1806	-0.0343	-0.0059	-0.1842	-0.4568	0.0287	0.0136
-1	-0.0621	-0.0621	-0.272	0	-0.3336	0.3336	0	-0.6166	0.1806	-0.0343	-0.0059	0.1842	0.4568	-0.0287	-0.0136
0.5699	0.4472	0.4472	1	0	0	0	0	0.6838	0.0296	0.695	1.1831	0	0	0	0
lamda matrix (real)				lamda matrix (imag)				vecTinv * vecT (real)				vecTinv * vecT (imag)			
-0.5699	0	0	0	0	0	0	0	1	0	0	0	0	0	0	0
0	-0.2412	0	0	0	1.2953	0	0	0	1	0	0	0	0	0	0
0	0	-0.2412	0	0	0	-1.2953	0	0	0	1	0	0	0	0	0
0	0	0	-3.6765	0	0	0	0	0	0	0	1	0	0	0	0
Ac	Bc	WR	WI											
-0.0706	0	-0.5699	0											
0.0027	0.0311	-0.2412	1.2953											
0.0027	-0.0311	-0.2412	-1.2953											
-0.0067	0	-3.6765	0											

The program now outputs the A_C and B_C coefficients for rudder response. These are input to Y:

$$Y_{Dr} = -0.071e^{-0.57t} + 0.071$$
$$+ 2\{e^{-0.24t}[0.0027\cos(1.3t) - 0.0311\sin(1.3t)] - 0.0027\}$$
$$- 0.0067e^{-3.68t} + 0.0067$$

The following is a plot of Y_{Dr}. Comparison with Runge-Kutta shows that Y_{Dr} is correct.

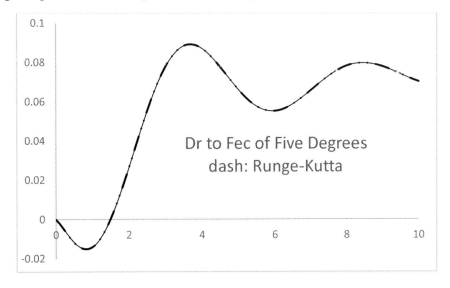

Dr to Fec of Five Degrees
dash: Runge-Kutta

Applications

Case 2: Balancing a Double Inverted Pendulum on a Cart

The schematic is shown in the following figure.

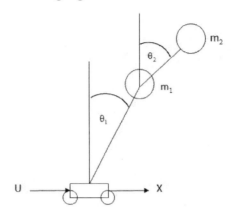

The system equations are shown in the following table.

The System Equations	
Motion Equations	Control System Equations
$\ddot{\theta}_1 = b_1\theta_1 - b_2\theta_2 + b_3 U$ $\ddot{\theta}_2 = -b_4\theta_1 + b_5\theta_2 + b_6 U$ $\ddot{X} = b_8\theta_1 + b_9\theta_2 + b_7 U$	$U = K_1(X_C - X) - K_2\dot{X} - K_3\theta_1 - K_4\dot{\theta}_1 - K_5\theta_2 - K_6\dot{\theta}_2$ where X_C is the cart position-change command

The motion equations and their data are from Mitchell and Gauthier Associates.[12] The control system is a personal design. The A and B matrices are:

$$\begin{bmatrix} \dot{\theta}_1 \\ \dot{\theta}_2 \\ \ddot{\theta}_1 \\ \ddot{\theta}_2 \\ \dot{X} \\ \ddot{X} \end{bmatrix} = \begin{bmatrix} 0 & 0 & 1 & 0 & 0 & 0 \\ 0 & 0 & 0 & 1 & 0 & 0 \\ -80.2 & 89.5 & -5.26 & 9.2 & 0.126 & 0.36 \\ -22.9 & 20.7 & 1.25 & -2.2 & -0.03 & -0.087 \\ 0 & 0 & 0 & 0 & 0 & 1 \\ 50.8 & -44.6 & 2.04 & -3.6 & -0.05 & -0.14 \end{bmatrix} \begin{bmatrix} \theta_1 \\ \theta_2 \\ \dot{\theta}_1 \\ \dot{\theta}_2 \\ X \\ \dot{X} \end{bmatrix} + \begin{bmatrix} 0 \\ 0 \\ -0.0629 \\ 0.015 \\ 0 \\ 0.0244 \end{bmatrix} X_C.$$

The problem statement is to compute the eigenvalues and eigenvectors for the A matrix and to use them to compute responses due to $X_C = 0.5$. Note that the size of the command (0.5) is included in the B matrix.

---

[12] Mitchell and Gauthier Associates, *A Model of Balancing Two Stacked Vertical Sticks* (Concord, MA: 1989).

The following is the spreadsheet when the A matrix is input to Program *Sub eig_main*.

Spreadsheet for Program *Sub eig_main*							
6	"MG iDP"	sheet 24				WR(i)	WI(i)
0	0	1	0	0	0	-0.199	0.349
0	0	0	1	0	0	-0.199	-0.349
-80.2	89.5	-5.26	9.2	0.126	0.36	-1.181	2.747
-22.9	20.7	1.25	-2.2	-0.03	-0.087	-1.181	-2.747
0	0	0	0	0	1	-2.42	5.51
50.8	-44.6	2.04	-3.6	-0.05	-0.14	-2.42	-5.51
eigsum	trace	prod	deter				
-7.6	-7.6	52.368	52.369				

For the θ_1 mode of motion, $\lambda = -2.42 \pm j5.51$. For the θ_2 mode of motion, $\lambda = -1.18 \pm j2.75$. For the X mode of motion, $\lambda = -0.2 \pm j0.35$.

The following is the spreadsheet when the A matrix and its eigenvalues are input to Program *Sub eigvec_main*.

Spreadsheet for Program *Sub eigvec_main* for T1 (θ_1) Response to Xc = .5											
6	DIP	sheet 17		T1/Xc		WR(i)	WI(i)		"B"		"C"
0	0	1	0	0	0	-0.199	0.349		0		57.296
0	0	0	1	0	0	-0.199	-0.349		0		0
-80.2	89.5	-5.26	9.2	0.126	0.36	-1.181	2.747		-0.0629		0
-22.9	20.7	1.25	-2.2	-0.03	-0.087	-1.181	-2.747		0.015		0
0	0	0	0	0	1	-2.42	5.51		0		0
50.8	-44.6	2.04	-3.6	-0.05	-0.14	-2.42	-5.51		0.0244		0
1	1	1	1	1	1	values of op for irow = 1					
1	1	1	1	1	1	values of pass for irow = 1					
eigenvectors (real)						eigenvectors (imag)					
-0.008	-0.008	-0.0282	-0.0282	0.0108	0.0108	0.0142	-0.0142	0.2044	-0.2044	0.1658	-0.1658
-0.0079	-0.0079	0.0059	0.0059	0.0494	0.0494	0.0142	-0.0142	0.1721	-0.1721	0.0693	-0.0693
-0.0034	-0.0034	-0.5281	-0.5281	-0.9397	-0.9397	-0.0057	0.0057	-0.3189	0.3189	-0.3419	0.3419
-0.0034	-0.0034	-0.4796	-0.4796	-0.5015	-0.5015	-0.0056	0.0056	-0.1871	0.1871	0.1044	-0.1044
-0.4953	-0.4953	-0.1321	-0.1321	-0.0517	-0.0517	-0.8687	0.8687	-0.3072	0.3072	-0.1178	0.1178
0.4017	0.4017	1	1	0.774	0.774	0	0	0	0	0	0

The lamda matrix is not shown, but the maximum size of the off-diagonal terms is 0.0006.

The B and C matrices for θ_1 (or T1) have also been input. The following table shows the resulting coefficients.

Step-Response Coefficients from Program *Sub eigvec_main* for T1 Response to Xc = .5			
Ac	Bc	WR	WI
0.2701	0.1145	-0.199	0.3494
0.2701	-0.1145	-0.199	-0.3494
-0.4075	0.1864	-1.1814	2.7474
-0.4075	-0.1864	-1.1814	-2.7474
0.1356	-0.0821	-2.4196	5.5098
0.1356	0.0821	-2.4196	-5.5098

These coefficients form the following equation:

$$T_1 = 2\{e^{-0.2t}[\ 0.2701\cos(0.35t) - 0.115\sin(0.35t)] - 0.2701\}$$
$$+ 2\{e^{-1.18t}[-0.408\cos(2.75t) - 0.186\sin(2.75t)] + 0.408\}$$
$$+ 2\{e^{-2.42t}[\ 0.136\cos(5.5t) + 0.082\sin(5.5t)] - 0.136\}$$

The following is the plot. The response is compared with Runge-Kutta integration of the system equations. The data used in the Runge-Kutta program are in appendix E.

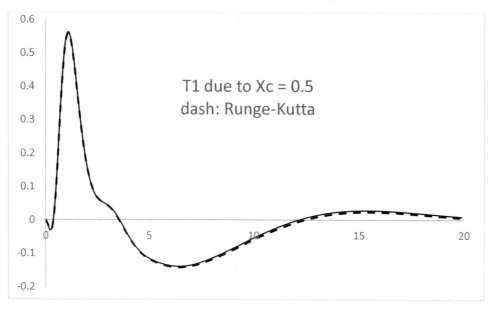

Another computer run was made to get the response of θ_2 (or T2). The B matrix does not change. Only the C matrix is changed. The following table shows the C matrix and the resulting coefficients for θ_2.

Step-Response Coefficients from Program Sub eigvec_main for T2 Response to Xc = .5				
C Matrix	Coefficients			
	Ac	Bc	WR	WI
0	0.2698	0.1129	-0.199	0.3494
57.296	0.2698	-0.1129	-0.199	-0.3494
0	-0.3085	0.2113	-1.1814	2.7474
0	-0.3085	-0.2113	-1.1814	-2.7474
0	0.0369	-0.0723	-2.4196	5.5098
0	0.0369	0.0723	-2.4196	-5.5098

These coefficients form the following equation:

$$T_2 = 2\{e^{-0.2t}[\,0.27\cos(0.35t) - 0.113\sin(0.35t)] - 0.27\}$$
$$+ 2\{e^{-1.18t}[-0.309\cos(2.75t) - 0.211\sin(2.75t)] + 0.309\}$$
$$+ 2\{e^{-2.42t}[\,0.037\cos(5.5t) + 0.072\sin(5.5t)] - 0.037\}$$

The following is the plot. The response is compared with Runge-Kutta integration of the system equations.

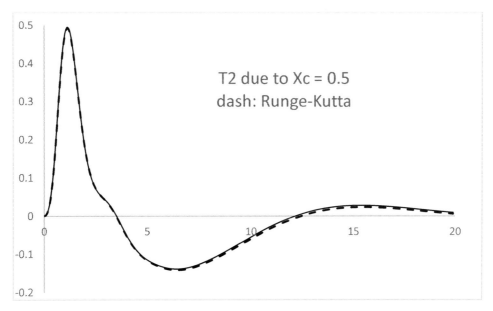

Another computer run was made to get the response of X. The B matrix does not change. Only the C matrix is changed. The following table shows the C matrix and the resulting coefficients for X.

Step-Response Coefficients from Program *Sub eigvec_main* for X Response to Xc = .5				
C Matrix	Coefficients			
	Ac	Bc	WR	WI
0	-0.2539	0.1858	-0.199	0.3494
0	-0.2539	-0.1858	-0.199	-0.3494
0	0.0071	-0.0105	-1.1814	2.7474
0	0.0071	0.0105	-1.1814	-2.7474
1	-0.0013	0.0017	-2.4196	5.5098
0	-0.0013	-0.0017	-2.4196	-5.5098

These coefficients form the following equation:

$$X = 2\{e^{-0.2t}[-0.254\cos(0.35t) - 0.186\sin(0.35t)] + 0.254\}$$
$$+ 2\{e^{-1.18t}[0.007\cos(2.75t) + 0.01\sin(2.75t)] - 0.007\}$$
$$+ 2\{e^{-2.42t}[-0.0013\cos(5.5t) - 0.0017\sin(5.5t)] + 0.0013\}$$

The following is the plot. The response is compared with Runge-Kutta integration of the system equations.

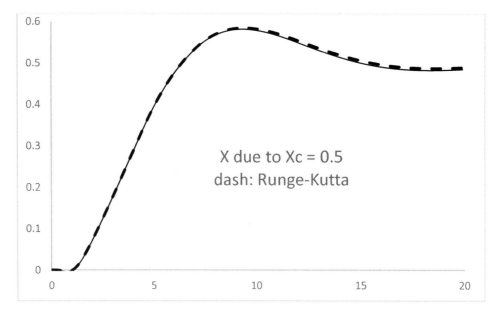

X due to Xc = 0.5
dash: Runge-Kutta

Applications

Case 3: Attitude Control of a Large, Flexible Launch Vehicle

The A matrix for this case comes from an actual math model of a large launch vehicle. This is one of the standard math models used in the design of a vehicle's flight-control system. Greensite is a good reference.[13] The equations are shown on the next page. The 14 state variables are defined there.

The A matrix is computed from data during the launch phase, where the aerodynamic pressure is the highest. This A matrix follows:

$$\begin{bmatrix} 0 & 1 & 0 & 0 & 0 & 0 & 0 & 0 & 0 & 0 & 0 & 0 & 0 & 0 \\ 3.5 & 0 & 0.00269 & 0 & 0 & 0 & 0 & 0 & 0 & -0.06 & 0 & 0 & 4.6 & 0 \\ -112.727 & 0 & -0.02797 & 0 & 0 & 0 & 0 & 0 & 0 & 0 & 0 & 0 & 61.8182 & 0 \\ 0 & 0 & 0 & 0 & 1 & 0 & 0 & 0 & 0 & 0 & 0 & 0 & 0 & 0 \\ 0 & 4900 & 0 & -4900 & -70 & 0 & 49 & 0 & 24.5 & 0 & 0 & 0 & 0 & 0 \\ 0 & 0 & 0 & 0 & 0 & 0 & 1 & 0 & 0 & 0 & 0 & 0 & 0 & 0 \\ 0 & 0 & 0 & 0 & 0 & -400 & -0.4 & 0 & 0 & 0 & 0 & 0 & -213.836 & 0 \\ 0 & 0 & 0 & 0 & 0 & 0 & 0 & 0 & 1 & 0 & 0 & 0 & 0 & 0 \\ 0 & 0 & 0 & 0 & 0 & 0 & 0 & -900 & -0.6 & 0 & 0 & 0 & -425 & 0 \\ 0 & 0 & 0 & 0 & 0 & 0 & 0 & 0 & 0 & 0 & 1 & 0 & 0 & 0 \\ 0 & 0 & 0 & 0 & 0 & 0 & 0 & 0 & 0 & -26.01 & -0.102 & 0 & -17.9 & 0 \\ 0 & 0 & 0 & -2520 & 0 & 0 & 0 & 0 & 0 & 0 & 0 & -84 & -3600 & -7560 \\ 0 & 0 & 0 & 0 & 0 & 0 & 0 & 0 & 0 & 0 & 0 & 1 & 0 & 0 \\ 35 & 0 & 0 & 0 & 0 & 0.35 & 0 & 0.175 & 0 & 0 & 0 & 0 & 0 & -35 \end{bmatrix} \begin{bmatrix} \theta \\ \dot\theta \\ \dot z \\ \dot\theta_f \\ \ddot\theta_f \\ q_1 \\ \dot q_1 \\ q_2 \\ \dot q_2 \\ \Gamma \\ \dot\Gamma \\ \dot\delta \\ \delta \\ \theta_f \end{bmatrix}$$

The task is to compute the eigenvalues and eigenvectors of this matrix. Then use these to compute the vehicle's attitude response to a commanded change of one degree ($\theta_c = 1$). To make things interesting, adjustments have been made to the autopilot, causing the first structural bending mode to be unstable.

---

[13] A. Greensite, *Analysis and Design of Space Vehicle Flight Control Systems* (New York: Spartan Books, 1970).

Math Model for the Design of Flight Control Systems for Large Launch Vehicles

Motions and Control System Equations	State Variables
Attitude and rate sensor outputs $\theta_s = \theta + \sigma_1 * q_1 + \sigma_2 * q_2$ $\dot{\theta}_s = \dot{\theta} + \sigma_1 * \dot{q}_1 + \sigma_2 * \dot{q}_2$ **Attitude filter output in the autopilot** $\dfrac{d}{dt}\theta_f = \omega_D(\theta_s - \theta_f)$ **Rate filter output in the autopilot** $\dfrac{d}{dt}(\ddot{\theta}_f) = \omega_R^2(\dot{\theta}_s - \dot{\theta}_f) - 2\zeta_R \omega_R \ddot{\theta}_f$ **Autopilot command** $\delta_c = k_D(\theta_c - \theta_f) - k_R \dot{\theta}_f$ **Engine angle acceleration** $\ddot{\delta} = \omega_a^2(\delta_c - \delta) - 2\zeta_a \omega_a \dot{\delta}$ **Vehicle angle of attack** $\alpha = \theta + \dot{Z}/V$ **Vehicle angular acceleration** $\ddot{\theta} = \mu_\alpha \alpha + \mu_\delta \delta - \mu_\Gamma \Gamma$ **Vehicle normal acceleration** $\ddot{Z} = -\dfrac{T-D}{m}\theta - \dfrac{L_\alpha}{m}\alpha + \dfrac{T_C}{m}\delta$ **First bending mode acceleration** $\ddot{q}_1 = -\dfrac{\partial \ddot{q}_1}{\partial \delta}\delta - 2\zeta_{B1}\omega_{B1} * \dot{q}_1 - \omega_{B1}^2 * q_1$ **Second bending mode acceleration** $\ddot{q}_2 = -\dfrac{\partial \ddot{q}_2}{\partial \delta}\delta - 2\zeta_{B2}\omega_{B2} * \dot{q}_2 - \omega_{B2}^2 * q_2$ **Fuel slosh mode acceleration** $\ddot{\Gamma} = -\dfrac{\partial \ddot{\Gamma}}{\partial \delta}\delta - 2\zeta_\Gamma \omega_\Gamma * \dot{\Gamma} - \omega_\Gamma^2 * \Gamma$	1. θ vehicle attitude 2. $\dot{\theta}$ vehicle rate 3. $\dot{Z}$ vehicle normal acceleration 4. $\dot{\theta}_f$ rate filter output 5. $\ddot{\theta}_f$ rate filter, internal variable 6. q_1 first bending mode displacement 7. $\dot{q}_1$ first bending mode rate 8. q_2 second bending mode displacement 9. $\dot{q}_2$ second bending mode rate 10. Γ fuel slosh angle 11. $\dot{\Gamma}$ fuel slosh angular rate 12. $\dot{\delta}$ engine angular rate 13. δ engine angle 14. θ_f attitude filter output

The following table shows the eigenvalues computed by program *Sub eig_main*.

Eigenvalues Computed by Program *Sub eig_main*		
WR	WI	Source of Each Eigenvalue
-0.1346	0	Three from the rigid-body stabilized by the control system
-1.5168	2.1407	
-1.5168	-2.1407	
-0.0315	5.0992	Two from fuel-slosh
-0.0315	-5.0992	
0.4503	19.4115	Two from the first bending-mode
0.4503	-19.4115	
-0.0641	29.2568	Two from the second bending-mode
-0.0641	-29.2568	
-35.2156	0	One from the attitude-loop-filter
-41.1262	43.4577	Two from the actuator
-41.1262	-43.4577	
-35.1016	60.2154	Two from the rate-loop-filter
-35.1016	-60.2154	

The printout showed that the eigenvalues are accurate because **eigsum = trace** and **prod = deter**. As expected, the first bending mode is unstable.

The A matrix and its eigenvalues are then input to program *Sub eigvec_main*. In addition, the B matrix is input for θ_c, and the C matrix is input for θ_f. The resulting eigenvectors are shown in appendix F.

Appendix F also shows the *lamda* matrix. The largest off-diagonal terms are less than 0.0314. To see if this is accurate enough, the attitude response must be computed and plotted. The data to do this are shown in the following table.

Input Data and Results from Program Sub eigvec_main					
The B Matrix for θ_c and the C Matrix for θ_f		The Coefficients and the Eigenvalues			
B	C	Ac	Bc	WR	WI
0	0	0.9085	0	-0.1346	0
0	0	-0.8335	0.674	-1.5168	2.1407
0	0	-0.8335	-0.674	-1.5168	-2.1407
0	0	0.0011	0.0025	-0.0315	5.0992
0	0	0.0011	-0.0025	-0.0315	-5.0992
0	0	0.0056	-0.0038	0.4503	19.412
0	0	0.0056	0.0038	0.4503	-19.412
0	0	0.0003	-0.0019	-0.0641	29.257
0	0	0.0003	0.0019	-0.0641	-29.257
0	0	-0.006	0	-35.216	0
0	0	-0.0003	-0.0001	-41.126	43.458
7560	0	-0.0003	0.0001	-41.126	-43.458
0	0	0	0	-35.102	60.215
0	1	0	0	-35.102	-60.215

The coefficients and the eigenvalues are input to the following equation.

$$Y_\theta = 0.9085 e^{-0.135t} - 0.9085$$
$$+ 2\{e^{-1.52t}[-0.8335\cos(2.14t) - 0.6740\sin(2.14t)] + 0.8335\}$$
$$+ 2\{e^{-0.03t}[0.0011\cos(5.1t) - 0.0025\sin(5.1t)] - 0.0011\}$$
$$+ 2\{e^{+0.45t}[0.0056\cos(19.41t) + 0.0038\sin(19.41t)] - 0.0056\}$$
$$+ 2\{e^{-0.06t}[0.0003\cos(29.26t) + 0.0019\sin(29.26t)] - 0.0003\}$$
$$- 0.006 e^{-35.22t} + 0.006$$
$$+ 2\{e^{-41.1t}[-0.0003\cos(43.46t) + 0.0001\sin(43.46t)] + 0.0003\}$$

The following is the plot of this equation.

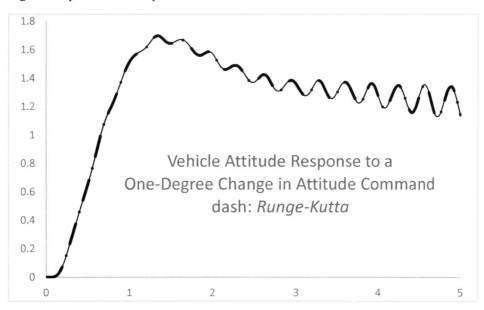

For checkout, the system equations were also integrated using the Runge-Kutta method programmed in chapter 11. The data to do this are given in appendix E. Comparison shows that the eigenvectors are accurate.

Applications

Case 4: Mode Shapes of a Free Vibrating System

Given the following spring-mass system, which has no damping:

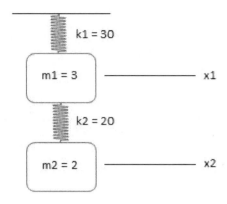

Using eigenvectors, determine the motions of each mass, if m_2 is moved up a distance of one unit. In other words, determine the motions if the initial conditions are $x_2 = 1$ and $x_1 = 0$.

Because there are two masses, there will be two response frequencies. Because there are no initial conditions on rate, the motions will have the following form:

$$x_1 = D_1 \cos(\omega_1 t) + D_2 \cos(\omega_2 t)$$
$$x_2 = D_3 \cos(\omega_1 t) + D_4 \cos(\omega_2 t)$$

Equation 1

We will use eigenvectors to find the values of the four coefficients.

The system equations are:

$$\begin{bmatrix} \ddot{x}_1 \\ \ddot{x}_2 \end{bmatrix} = \begin{bmatrix} -\dfrac{K_1+K_2}{m_1} & \dfrac{K_2}{m_1} \\ \dfrac{K_2}{m_2} & -\dfrac{K_2}{m_2} \end{bmatrix} \begin{bmatrix} x_1 \\ x_2 \end{bmatrix} = \begin{bmatrix} -16.667 & 6.667 \\ 10 & -10 \end{bmatrix} \begin{bmatrix} x_1 \\ x_2 \end{bmatrix}$$

Let's discuss the motion at a general frequency ω_i:

$$x_1 = \alpha \cos(\omega_i t) \text{ and } x_2 = \beta \cos(\omega_i t).$$

Equation 2

Taking derivatives: $\dot{x}_1 = -\omega_i \alpha \sin(\omega_i t)$ and $\dot{x}_2 = -\omega_i \beta \sin(\omega_i t)$.

Another derivative: $\ddot{x}_1 = -\omega_i^2 \alpha \cos(\omega_i t)$ and $\ddot{x}_2 = -\omega_i^2 \beta \cos(\omega_i t)$.

Substituting these motions and their second derivatives into the system equations yields:

$$\begin{bmatrix} -\omega_i^2 \alpha \\ -\omega_i^2 \beta \end{bmatrix} \cos(\omega_i t) = \begin{bmatrix} -16.667 & 6.667 \\ 10 & -10 \end{bmatrix} \begin{bmatrix} \alpha \\ \beta \end{bmatrix} \cos(\omega_i t).$$

This means that:
$$\omega_i^2 \begin{bmatrix} \alpha \\ \beta \end{bmatrix} = \begin{bmatrix} 16.667 & -6.667 \\ -10 & 10 \end{bmatrix} \begin{bmatrix} \alpha \\ \beta \end{bmatrix}.$$

Now, if we let $\lambda = \omega_i^2$, the last equation becomes:

$$\lambda \begin{bmatrix} \alpha \\ \beta \end{bmatrix} = \begin{bmatrix} 16.667 & -6.667 \\ -10 & 10 \end{bmatrix} \begin{bmatrix} \alpha \\ \beta \end{bmatrix} = [A] \begin{bmatrix} \alpha \\ \beta \end{bmatrix}.$$
Equation 3

Equation 3 has the form of the *Eigenvalue Problem*. The eigenvalues are the square of the frequencies for Equation 2. The components of the eigenvector are the coefficients for Equation 2.

When we put the A matrix of Equation 3 into program Sub eig_main, we get the following spreadsheet.

Spreadsheet for Program Sub eig_main				
2			WR(i)	WI(i)
16.66667	-6.66667		22.1525	0
-10	10		4.514162	0
eigsum	trace	prod	deter	
26.667	26.667	100	100	

There are two eigenvalues: $\lambda_1 = \omega_1^2 = 22.15 = 4.71^2$ and $\lambda_2 = \omega_2^2 = 4.51 = 2.12^2$.

When we put the A matrix and its eigenvalues into program Sub eigvec_main, we get the eigenvectors.

Spreadsheet for Program Sub eigvec_main								
2		WR(i)	WI(i)				lamda diagonal	
16.66667	-6.66667	22.1525	0				22.1525	0
-10	10	4.514162	0				4.514162	0
1	1	values of op for irow = 1						
1	1	values of pass for irow = 1						
eigenvectors (real)		eigenvectors (imag)			eigenvector matrix inverse (re	eigenvector mat inv(imag)		
-1	0.5486	0	0		-0.689	0.378	0	0
0.8229	1	0	0		0.5669	0.689	0	0
lamda matrix (real)		lamda matrix (imag)			vecTinv * vecT (real)		vecTinv * vecT (imag)	
22.1525	0	0	0		1	0	0	0
0	4.5142	0	0		0	1	0	0

The eigenvector matrix is: $\begin{bmatrix} -1 & 0.5486 \\ 0.8229 & 1 \end{bmatrix}$.

From the definition of eigenvectors: $\begin{bmatrix} x_1 \\ x_2 \end{bmatrix} = V \begin{bmatrix} q_1 \\ q_2 \end{bmatrix} = \begin{bmatrix} -1 & 0.5486 \\ 0.8229 & 1 \end{bmatrix} \begin{bmatrix} q_1 \\ q_2 \end{bmatrix}$, where q_1 is a function of λ_1 and q_2 is a function of λ_2.

Let's look at V as a *matrix of columns*. At each λ, it shows the motion of x_2 relative to x_1.

Therefore, let's normalize as follows: $\begin{bmatrix} x_1 \\ x_2 \end{bmatrix} = \begin{bmatrix} 1 & 1 \\ -0.8229 & 1.8228 \end{bmatrix} \begin{bmatrix} q_1 \\ q_2 \end{bmatrix}$.

This means that

- the motion of x_2 at ω_1 is (-0.8229) times the motion of x_1; and
- the motion of x_2 at ω_2 is (1.8228) times the motion of x_1.

Equation 1 can now be written:

$$x_1 = D_1 \cos(4.71t) + D_2 \cos(2.12t)$$
$$x_2 = -0.8229 D_1 \cos(4.71t) + 1.8228 D_2 \cos(2.12t)$$

At the initial conditions, t = 0. Therefore:

$$x_1 = 0 = D_1 + D_2$$
$$x_2 = 1 = -0.8229 D_1 + 1.8228 D_2$$

These equations are satisfied when $D_1 = -0.378$ and $D_2 = 0.378$.

Therefore:
$$x_1 = -0.378 \cos(4.71t) + 0.378 \cos(2.12t)$$
$$x_2 = 0.311 \cos(4.71t) + 0.689 \cos(2.12t)$$

These two equations are plotted in the following two figures.

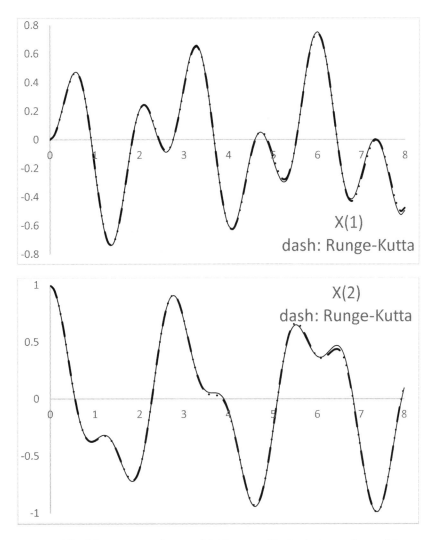

These equations are verified by comparison with Runge-Kutta integration of the system equations. The data used in the Runge-Kutta program are listed in appendix E.

This way of using eigenvectors is widely used in the field of structural dynamics. In that field the eigenvectors are called mode shapes.

Chapter 16: More Examples

These examples discuss situations that may be encountered when computing eigenvectors.

Example 16.1: Computing the Eigenvectors Iteratively
- V^{-1} exists, but L has large off-diagonal terms.

Example 16.2: On Drawing a Block Diagram of the A Matrix
- V^{-1} doesn't exist, and V is inaccurate.

Example 16.3: Defective Matrices
- V^{-1} doesn't exist, but V is accurate.

Example 16.4: Similar Matrices and Their Eigenvectors
- Here we can rearrange the A matrix to compute accurate eigenvalues and eigenvectors.

Example 16.5: Pole-zero Cancellation
- This example shows that eigenvectors can detect pole-zero cancellation.

More Examples

Example 16.1: Computing the Eigenvectors Iteratively

Computing eigenvalues for systems higher than second or third order is an iterative process. As such, this means that eigenvalues and eigenvectors can only be reasonably accurate. The procedure used in this example improves the accuracy of both.

We've been using the following notation:

- A is the system matrix.
- V is the eigenvector matrix.
- L = V⁻¹AV is the lamda matrix.

If L is a diagonal matrix, V is accurate. But practically speaking, there are always nonzero off-diagonal terms. The off-diagonal terms can be made smaller if the diagonal terms are used to recompute V. This example shows this process:

Given:
$$A = \begin{bmatrix} 0 & 1 & 0 \\ -1 & -1 & 1 \\ -0.5 & 0 & -0.5 \end{bmatrix}, B = \begin{bmatrix} 0 \\ 0 \\ 1 \end{bmatrix}, \text{ and } C = \begin{bmatrix} 1 & 0 & 0 \end{bmatrix}.$$

The following table is the spreadsheet showing the *good* eigenvalues.

The Spreadsheet for Program *Sub eig_main*					
3				WR(i)	WI(i)
0	1	0		-1	0
-1	-1	1		-0.25	0.968246
-0.5	0	-0.5		-0.25	-0.96825
eigsum	trace	prod	deter		
-1.5	-1.5	-1	-1		

Arbitrarily, let's compute V using the following eigenvalues:

$$\lambda's = -0.5, \; -0.5 + j0.866, \; -0.5 - j0.866.$$

The following table shows the results.

125

The Spreadsheet for Program Sub eigvec_main

3			start		WR(i)	WI(i)					lamda diagonal	
0	1	0			-0.5	0		0		1	-1.16667	1.67E-16
-1	-1	1			-0.5	0.866025		0		0	-0.16667	1.122626
-0.5	0	-0.5			-0.5	-0.86603		1		0	-0.16667	-1.12263
	1	1	1 values of op for irow = 1									
	1	1	1 values of pass for irow = 1									
	1	1	1 values of op for irow=2									
	1	2	2 values of pass for irow=2									

eigenvectors (real)			eigenvectors (imag)			eigenvector matrix inverse (real)			eigenvector mat inv(imag)		
1	0	0	0	-1	1	-0.4444	-0.8889	1.3333	0	0	0
-0.5	0.866	0.866	0	0.5	-0.5	0.2887	0.5774	0	0.7222	0.4444	-0.6667
0.75	0.5774	0.5774	0	0	0	0.2887	0.5774	0	-0.7222	-0.4444	0.6667

lamda matrix (real)			lamda matrix (imag)			vecTinv * vecT (real)			vecTinv * vecT (imag)		
-1.1667	-0.5132	-0.5132	0	0	0	1	0	0	0	0	0
0	-0.1667	0.3333	0.3333	1.1226	0.2566	0	1	0	0	0	0
0	0.3333	-0.1667	-0.3333	-0.2566	-1.1226	0	0	1	0	0	0

Ac	Bc	WR	WI
-1.1429	0	-1.1667	0
0.0863	0.581	-0.1667	1.1226
0.0863	-0.581	-0.1667	-1.1226

The lamda matrix has large off-diagonal terms. The following plot uses the above step-response coefficients.

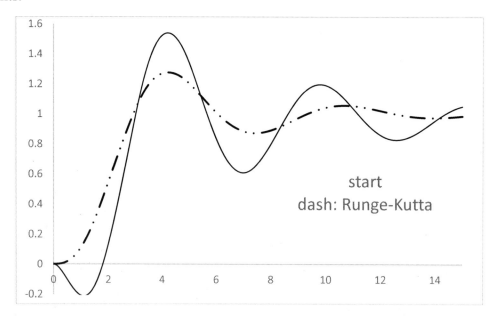

start

dash: Runge-Kutta

Compared with the results from Runge-Kutta, the error is large.

The diagonal terms of the lamda matrix are printed in the upper-right corner of the spreadsheet. These diagonal terms are then input as eigenvalues for a second iteration. The results are shown in the following table.

The Spreadsheet for the Second Run of Program Sub eigvec_main

3		two		WR(i)	WI(i)					lamda diagonal	
0	1	0		-1.16667	1.67E-16		0		1	-1.03558	2.39E-15
-1	-1	1		-0.16667	1.122626		0		0	-0.23221	0.985428
-0.5	0	-0.5		-0.16667	-1.12263		1		0	-0.23221	-0.98543
1	1	1	values of op for irow = 1								
1	1	1	values of pass for irow = 1								
1	1	1	values of op for irow=2								
1	1	1	values of pass for irow=2								

eigenvectors (real)			eigenvectors (imag)			eigenvector matrix inverse (real)			eigenvector mat inv(imag)		
0.8372	-0.4147	-0.4147	0	-0.7774	0.7774	0.1522	-0.3523	0.5284	0	0	0
-0.9767	0.9419	0.9419	0	-0.3359	0.3359	-0.1018	0.2357	0.3155	0.5069	0.3154	-0.1163
1	0.7474	0.7474	0	0	0	-0.1018	0.2357	0.3155	-0.5069	-0.3154	0.1163

lamda matrix (real)			lamda matrix (imag)			vecTinv * vecT (real)			vecTinv * vecT (imag)		
-1.0356	-0.0221	-0.0221	0	-0.238	0.238	1	0	0	0	0	0
0.0783	-0.2322	0.0392	-0.0288	0.9854	0.1469	0	1	0	0	0	0
0.0783	0.0392	-0.2322	0.0288	-0.1469	-0.9854	0	0	1	0	0	0

Ac	Bc	WR	WI
-0.4272	0	-1.0356	0
-0.1393	0.2573	-0.2322	0.9854
-0.1393	-0.2573	-0.2322	-0.9854

The off-diagonal terms of the lamda matrix are smaller. The following plot shows that the errors are smaller.

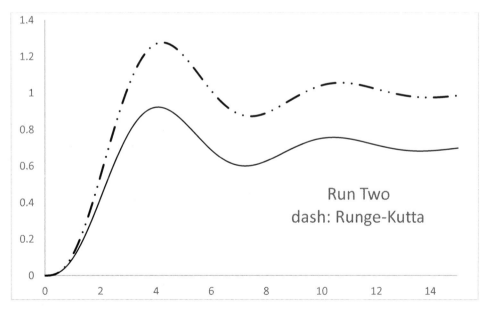

Run Two
dash: Runge-Kutta

After another iteration, the following table shows what happens when the third-iteration results are used as input for a fourth iteration.

The Spreadsheet for the Fourth Run of Program Sub eigvec_main

3		four		lamda diagonal						lamda diagonal		
0	1	0		-1.00136	1.51E-15		0		1	-1	-4.9E-16	
-1	-1	1		-0.24932	0.968469		0		0	-0.25	0.968246	
-0.5	0	-0.5		-0.24932	-0.96847		1		0	-0.25	-0.96825	
1	1	1	values of op for irow = 1									
1	1	1	values of pass for irow = 1									
1	1	1	values of op for irow=2									
1	1	1	values of pass for irow=2									
eigenvectors (real)				eigenvectors (imag)			eigenvector matrix inverse (real)			eigenvector mat inv(imag)		
0.9986	-0.2495	-0.2495		0	-0.9683	0.9683	0.0001	-0.3339	0.666	0	0	0
-1	1	1		0	-0.0002	0.0002	-0.0001	0.333	0.3331	0.5163	0.258	-0.2576
1	0.5014	0.5014		0	0	0	-0.0001	0.333	0.3331	-0.5163	-0.258	0.2576
lamda matrix (real)				lamda matrix (imag)			vecTinv * vecT (real)			vecTinv * vecT (imag)		
-1	-0.0006	-0.0006		0	-0.0009	0.0009	1	0	0	0	0	0
0.0007	-0.25	0		-0.0005	0.9682	0.0007	0	1	0	0	0	0
0.0007	0	-0.25		0.0005	-0.0007	-0.9682	0	0	1	0	0	0
Ac	Bc	WR	WI									
-0.6651	0	-1	0									
-0.1669	0.3866	-0.25	0.9682									
-0.1669	-0.3866	-0.25	-0.9682									

The lamda matrix is essentially diagonal. The following plot shows that the error is gone.

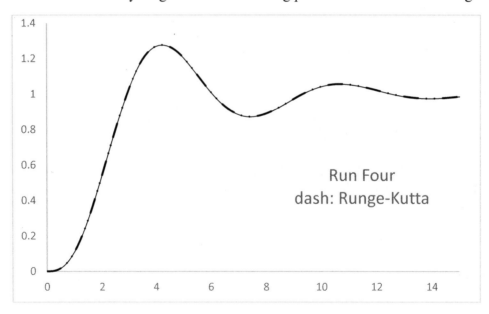

Run Four
dash: Runge-Kutta

This process won't turn complex eigenvalues into real eigenvalues, or vice versa. But it usually works.

Note: In program **Sub eigvec_main**, the outer loop indicated by the **irow** flag was needed for this A matrix.

Example 16.2: On Drawing a Block Diagram of the A Matrix

Practically speaking, there are always errors. Are the errors from the eigenvalues, from the eigenvectors, or from the A matrix itself?

At the risk of oversimplifying, consider the following:

$$A = \begin{bmatrix} -6 & 6 & 0 \\ 0 & -4 & 4 \\ 0 & 0 & -2 \end{bmatrix}, B = \begin{bmatrix} 0 \\ 0 \\ 2 \end{bmatrix}, \text{ and } C = \begin{bmatrix} 1 & 0 & 0 \end{bmatrix}.$$

By inspection, the eigenvalues are found to be -6, -4, and -2. But suppose we use -6.005, -4.005, and -2.005.

The following table shows inputting these into program Sub eigvec_main.

The Spreadsheet from Program Sub eigvec_main									
3 connected								lamda diagonal	
-6	6	0	-2.005	0	0	1	-2.00002	0	
0	-4	4	-4.005	0	0	0	-4	0	
0	0	-2	-6.005	0	2	0	-5.99998	0	
1	1	1 values of op for irow = 1							
3	3	3 values of pass for irow = 1							
eigenvectors (real)			eigenvectors (imag)			eigenvector matrix inverse (real)		eigenvector mat inv(imag)	
1	-1	1	0	0	0	-0.0012	0.0075	3	0	0	0
0.6658	-0.3325	-0.0008	0	0	0	-0.005	-2.985	6	0	0	0
0.3321	0.0004	0.0004	0	0	0	0.9963	-2.9925	3	0	0	0
lamda matrix (real)			lamda matrix (imag)			vecTinv * vecT (real)			vecTinv * vecT (imag)		
-2	0.0025	0.005	0	0	0	1	0	0	0	0	0
0.01	-4	0.01	0	0	0	0	1	0	0	0	0
0.005	0.0025	-6	0	0	0	0	0	1	0	0	0
Ac	Bc	WR	WI						
-3	0	-2	0						
3	0	-4	0						
-1	0	-6	0						

The lamda matrix has small off-diagonal terms, and its diagonal contains the good eigenvalues. These results form the equation for the step response, which is plotted in the following figure.

$$Y = -3*(e^{-2t}-1) + 3*(e^{-4t}-1) - (e^{-6t}-1).$$

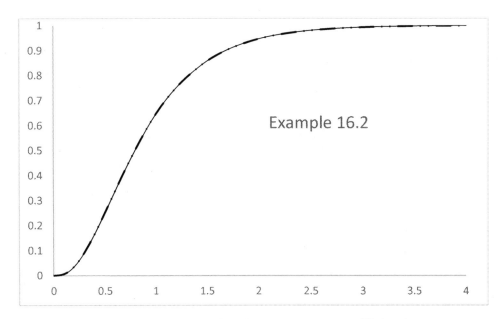

Example 16.2

Comparison with Runge-Kutta shows that the eigenvectors are sufficiently accurate.

But suppose the entry in row 1, column 2 of the A matrix was zero. The following table shows the results from program Sub eigvec_main.

The Spreadsheet from Program Sub eigvec_main												
3												
-6	0	0	-2.005	0		0		1				
0	-4	4	-4.005	0		0		0				
0	0	-2	-6.005	0		2		0				
1	1	1 values of op for irow = 1										
3	3	3 values of pass for irow = 1										
1	1	1 values of op for irow = 2. Hence, there is no vector inverse										
1	1	1 values of pass for irow = 2										
eigenvectors (real)			eigenvalues (imag)									
0	0	0	0	0	0							
1	-1	-1	0	0	0							
0.4988	0.0012	0.5013	0	0	0							
A*V (real)			A*V (imag)			V*L (real)			V*L (imag)			
0	0	0	0	0	0	0	0	0	0	0	0	
-2.005	4.005	6.005	0	0	0	-2.005	4.005	6.005	0	0	0	
-0.9975	-0.0025	-1.0025	0	0	0	-1	-0.005	-3.01	0	0	0	

V^{-1} could not be computed after trying all 54 combinations of equations and variables. When this happens, the program prints out two matrix products, AV and VL. The definition of eigenvectors is AV = VL. The printout shows that V is incorrect.

But suppose we put in the exact eigenvalues. The following table shows the results.

More Examples

The Spreadsheet from Program *Sub eigvec_main*									
3 unconnected							lamda diagonal		
-6	0	0	-2	0	0	1	-2	0	
0	-4	4	-4	0	0	0	-4	0	
0	0	-2	-6	0	2	0	-6	0	
1	2	3 values of op for irow = 1							
3	2	1 values of pass for irow = 1							
eigenvectors (real)			eigenvectors (imag)			eigenvector matrix inverse (real)			eigenvector mat inv(imag)
∞0	0	1	0	0	0	0	0	2	0	0	0
1	1	0	0	0	0	0	1	-2	0	0	0
0.5	0	0	0	0	0	1	0	0	0	0	0
lamda matrix (real)			lamda matrix (imag)			vecTinv * vecT (real)			vecTinv * vecT (imag)
-2	0	0	0	0	0	1	0	0	0	0	0
0	-4	0	0	0	0	0	1	0	0	0	0
0	0	-6	0	0	0	0	0	1	0	0	0
Ac	Bc	WR	WI						
0	0	-2	0						
0	0	-4	0						
0	0	-6	0						

V^{-1} exists, and good eigenvectors have been computed. But the step response shows zero output. Let's draw the block diagram of the A matrix from this last spreadsheet. This matrix can come from the following equations:

$$\dot{Z}_1 = -6 * Z_1$$
$$\dot{Z}_2 = -4 * Z_2 + 4 * Z_3$$
$$\dot{Z}_3 = -2 * Z_3 + 2 * U$$
$$Y = Z_1$$

The following is the block diagram of these equations.

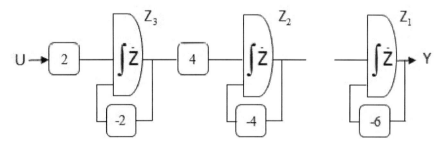

It shows that the A matrix has unconnected segments. The eigenvectors show this. But to find this out, we had to put in exact eigenvalues.

Problems often can't be detected from the matrix itself. A block diagram can be helpful. If the A matrix is meant to be unconnected, then find its eigenvectors from each segment separately.

More Examples

Example 16.3: Defective Matrices

A matrix is said to be defective if it has eigenvalues that are equal. As such, its matrix of eigenvectors cannot be inverted. This example shows how the programs in this book deal with these matrices.

Matrix 1: $\quad A = \begin{bmatrix} 0 & 1 & 1 \\ -2 & -3 & 2 \\ 0 & 0 & -2 \end{bmatrix}$.

To compute the eigenvalues, the matrix is input to program **Sub eig_main** as shown in the following table:

The spreadsheet for Program *Sub eig_main*					
3				WR(i)	WI(i)
0	1	1		-2	0
-2	-3	2		-1	0
0	0	-2		-2	0
eigsum	trace	prod	deter		
-5	-5	-4	-4		

There are two identical eigenvalues at λ = -2.

The following table shows the matrix and its eigenvalues input to program **Sub_eigvec_main**.

The Spreadsheet for Program *Sub eigvec_main*												
3			WR(i)	WI(i)								
0	1	1	-1	0								
-2	-3	2	-2	0								
0	0	-2	-2	0								
2	2	2 values of op for irow = 1										
1	3	3 values of pass for irow = 1										
2	2	2 values of op for irow = 2. Hence, there is no vector inverse										
1	1	1 values of pass for irow = 2										
eigenvectors (real)			eigenvalues (imag)									
-1	-0.5	-0.5	0	0	0							
1	1	1	0	0	0							
0	0	0	0	0	0							
A*V (real)			A*V (imag)			V*L (real)			V*L (imag)			
1	1	1	0	0	0	1	1	1	0	0	0
-1	-2	-2	0	0	0	-1	-2	-2	0	0	0
0	0	0	0	0	0	0	0	0	0	0	0

More Examples

As expected, the eigenvector matrix V cannot be inverted. The program prints out two matrix products, AV and VL. The definition of an eigenvector is AV = VL. Hence, the printout shows that the eigenvectors are accurate.

Matrix 2: $\quad A = \begin{bmatrix} 5 & 2 & 2 \\ 3 & 6 & 3 \\ 6 & 6 & 9 \end{bmatrix}$.

This matrix is input to program **Sub eig_main**:

The Spreadsheet for Program *Sub eig_main*					
3				WR(i)	WI(i)
5	2	2		14	0
3	6	3		3	0
6	6	9		3	0
eigsum	trace	prod	deter		
20	20	126	126		

There are two identical eigenvalues at $\lambda = 3$.

Before we put this matrix and its eigenvalues into program **Sub eigvec_main**, let's compute the eigenvectors by hand.

The **base equations** for the matrix are:
$$\begin{bmatrix} 5-\lambda & 2 & 2 \\ 3 & 6-\lambda & 3 \\ 6 & 6 & 9-\lambda \end{bmatrix} \begin{bmatrix} x_1 \\ x_2 \\ x_3 \end{bmatrix} = \begin{bmatrix} 0 \\ 0 \\ 0 \end{bmatrix}.$$

- For the eigenvalue $\lambda = 14$, the base equations become:
$$\begin{aligned} E1: &-9x_1 + 2x_2 + 2x_3 = 0 \\ E2: &3x_1 - 8x_2 + 3x_3 = 0 \\ E3: &6x_1 + 6x_2 - 5x_3 = 0 \end{aligned}$$

Pick $x_3 = 1$. Equations E2 and E3 become: $\begin{aligned} E2: & 3x_1 - 8x_2 = -3 \\ E3: & 6x_1 + 6x_2 = 5 \end{aligned}$. From these, compute $x_1 = 1/3$ and $x_2 = 1/2$. The eigenvector for $\lambda = 14$ is $\begin{bmatrix} 1/3 & 1/2 & 1 \end{bmatrix}^T$.

- For the eigenvalue $\lambda = 3$, the base equations become: $\begin{bmatrix} 2 & 2 & 2 \\ 3 & 3 & 3 \\ 6 & 6 & 6 \end{bmatrix} \begin{bmatrix} x_1 \\ x_2 \\ x_3 \end{bmatrix} = \begin{bmatrix} 0 \\ 0 \\ 0 \end{bmatrix}$. To solve these, two values for x_i must be picked. If we pick $x_2 = x_3 = 1$, we can solve for $x_1 = -2$. We have chosen the eigenvector for $\lambda = 3$ to be $\begin{bmatrix} -2 & 1 & 1 \end{bmatrix}^T$.

Combining these eigenvectors: $V = \begin{bmatrix} -2 & -2 & 1/3 \\ 1 & 1 & 1/2 \\ 1 & 1 & 1 \end{bmatrix}$.

Now we compute: $AV = \begin{bmatrix} 5 & 2 & 2 \\ 3 & 6 & 3 \\ 6 & 6 & 9 \end{bmatrix} \begin{bmatrix} -2 & -2 & 1/3 \\ 1 & 1 & 1/2 \\ 1 & 1 & 1 \end{bmatrix} = \begin{bmatrix} -6 & -6 & 14/3 \\ 3 & 3 & 7 \\ 3 & 3 & 14 \end{bmatrix}$,

and: $VL = \begin{bmatrix} -2 & -2 & 1/3 \\ 1 & 1 & 1/2 \\ 1 & 1 & 1 \end{bmatrix} \begin{bmatrix} 3 & 0 & 0 \\ 0 & 3 & 0 \\ 0 & 0 & 14 \end{bmatrix} = \begin{bmatrix} -6 & -6 & 14/3 \\ 3 & 3 & 7 \\ 3 & 3 & 14 \end{bmatrix}$.

Since AV = VL, the eigenvectors are correct. Now that we know the answer, let's put the matrix and its eigenvalues into program Sub_eigvec_main.

The Spreadsheet for Program *Sub eigvec_main*					
3				WR(i)	WI(i)
5	2	2		14	0
3	6	3		3	0
6	6	9		3	0
Eigenvector could not be computed					
1					
1					
0.3333				0	
0.5				0	
1				0	

It correctly computes the eigenvector for $\lambda = 14$. But it is unable to compute the vector for $\lambda = 3$. This is because Sub eigvec_main is programmed to pick only one of the elements of each eigenvector.

Example 16.4: Similar Matrices and Their Eigenvectors

Matrices are said to be similar if they have the same eigenvalues. This example shows what happens to the eigenvectors. We will use the matrix from example 2 in chapter 1.

$$A = \begin{bmatrix} 0 & 1 & 0 \\ -2 & -0.5 & 2 \\ 0 & 0 & -10 \end{bmatrix}$$

The following system of differential equations has this A matrix:

$$\begin{bmatrix} \dot{Z}_1 \\ \dot{Z}_2 \\ \dot{Z}_3 \end{bmatrix} = \begin{bmatrix} 0 & 1 & 0 \\ -2 & -0.5 & 2 \\ 0 & 0 & -10 \end{bmatrix} \begin{bmatrix} Z_1 \\ Z_2 \\ Z_3 \end{bmatrix} + \begin{bmatrix} 0 \\ 0 \\ 10 \end{bmatrix} U \quad \text{and} \quad Y = Z_1$$

The following is a block diagram of these equations:

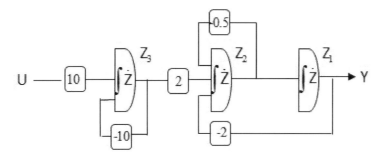

The following table is a repeat that shows the eigenvalues of the A matrix.

Spreadsheet for Program *Sub eig_main* for Example 2 Chapter 1					
3				WR(i)	WI(i)
0	1	0		-10	0
-2	-0.5	2		-0.25	1.391941
0	0	-10		-0.25	-1.39194
eigsum	trace	prod	deter		
-10.5	-10.5	-20	-20		

135

The following table is also a repeat showing the eigenvectors and the step-response coefficients for the system.

Spreadsheet for Program *Sub eigvec_main* for Example 2 Chapter 1											
3				WR(i)	WI(i)				lamda diagonal		
0	1	0		-10	0		0	1	-10	7.77E-15	
-2	-0.5	2		-0.25	1.391941		0	0	-0.25	1.391941	
0	0	-10		-0.25	-1.39194		10	0	-0.25	-1.39194	
	1	2	2 values of op for irow = 1								
	3	1	1 values of pass for irow = 1								
eigenvectors (real)				eigenvectors (imag)			eigenvector matrix inverse (real)		eigenvector mat inv(imag)		
0.0206	-0.125	-0.125		0	-0.696	0.696	0	0	1	0	0	0
-0.2062	1	1		0	0	0	0	0.5	0.1031	0.7184	0.0898	0.0037
1	0	0		0	0	0	0	0.5	0.1031	-0.7184	-0.0898	-0.0037
lamda matrix (real)				lamda matrix (imag)			vecTinv * vecT (real)		vecTinv * vecT (imag)		
-10	0	0		0	0	0	1	0	0	0	0	0
0	-0.25	0		0	1.3919	0	0	1	0	0	0	0
0	0	-0.25		0	0	-1.3919	0	0	1	0	0	0
Ac	Bc	WR	WI								
-0.0206	0	-10	0								
-0.4897	0.162	-0.25	1.3919								
-0.4897	-0.162	-0.25	-1.3919								

Let's renumber the output of the integrators as shown in the following block diagram:

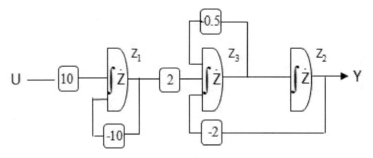

Renumbering causes matrix rows and columns to be interchanged. The renumbered system equations are:

$$\begin{bmatrix} \dot{Z}_1 \\ \dot{Z}_2 \\ \dot{Z}_3 \end{bmatrix} = \begin{bmatrix} -10 & 0 & 0 \\ 0 & 0 & 1 \\ 2 & -2 & -0.5 \end{bmatrix} \begin{bmatrix} Z_1 \\ Z_2 \\ Z_3 \end{bmatrix} + \begin{bmatrix} 10 \\ 0 \\ 0 \end{bmatrix} U \text{ and } Y = Z_2.$$

From these: $A = \begin{bmatrix} -10 & 0 & 0 \\ 0 & 0 & 1 \\ 2 & -2 & -0.5 \end{bmatrix}$, $B = \begin{bmatrix} 10 \\ 0 \\ 0 \end{bmatrix}$, and $C = \begin{bmatrix} 0 & 1 & 0 \end{bmatrix}$.

The following table shows the eigenvalues of this renumbered matrix.

Spreadsheet for Program *Sub eig_main* for Example 2 Chapter 1 Renumbered					
3				WR(i)	WI(i)
-10	0	0		-0.25	1.391941
0	0	1		-0.25	-1.39194
2	-2	-0.5		-10	0
eigsum	trace	prod	deter		
-10.5	-10.5	-20	-20		

The eigenvalues are the same as those from the original A matrix, showing that the matrices are similar.

The following table shows the eigenvectors and the step-response coefficients for this renumbered system.

Spreadsheet for Program *Sub eigvec_main* for Example 2 Chapter 1 Renumbered											
3			WR(i)	WI(i)					lamda diagonal		
-10	0	0	-0.25	1.391941		10	0	0	-0.25	1.391941	
0	0	1	-0.25	-1.39194		0	1	0	-0.25	-1.39194	
2	-2	-0.5	-10	0		0	0	0	-10	0	
1	1	1	values of op for irow = 1								
1	1	1	values of pass for irow = 1								
eigenvectors (real)			eigenvectors (imag)			eigenvector matrix inverse (real)			eigenvector mat inv(imag)		
0	0	-1	0	0	0	0.1031	0	0.5	0.0037	0.7184	0.0898
-0.125	-0.125	-0.0206	-0.696	0.696	0	0.1031	0	0.5	-0.0037	-0.7184	-0.0898
1	1	0.2062	0	0	0	-1	0	0	0	0	0
lamda matrix (real)			lamda matrix (imag)			vecTinv * vecT (real)			vecTinv * vecT (imag)		
-0.25	0	0	1.3919	0	0	1	0	0	0	0	0
0	-0.25	0	0	-1.3919	0	0	1	0	0	0	0
0	0	-10	0	0	0	0	0	1	0	0	0
Ac	Bc	WR	WI								
-0.4897	0.162	-0.25	1.3919								
-0.4897	-0.162	-0.25	-1.3919								
-0.0206	0	-10	0								

The eigenvectors are different, but the step-response coefficients are the same since the system didn't change.

Conclusion

Rearranging the numbers in a matrix by renumbering the state variables, creates an A matrix that is similar to the original. The response will be the same. Sometimes, this rearranging can improve the accuracy of the eigenvectors.

Example 16.5: Pole-Zero Cancellation

Given these equations: $\begin{bmatrix} \dot{Z}_1 \\ \dot{Z}_2 \\ \dot{Z}_3 \end{bmatrix} = \begin{bmatrix} -2 & 1 & 1 \\ -3 & 1 & 2 \\ 0 & 0 & -1 \end{bmatrix} \begin{bmatrix} Z_1 \\ Z_2 \\ Z_3 \end{bmatrix} + \begin{bmatrix} 0 \\ 0 \\ 1 \end{bmatrix} U$ and $Y = Z_1$.

The following table shows the eigenvalues of the A matrix.

The Spreadsheet for Program Sub eig_main					
3				WR(i)	WI(i)
-2	1	1		-1	0
-3	1	2		-0.5	0.866025
0	0	-1		-0.5	-0.86603
eigsum	trace	prod	deter		
-2	-2	-1	-1		

The following table shows the eigenvectors and the step-response coefficients.

The Spreadsheet for Program Sub eigvec_main											
3	pole-zero cancel		WR(i)	WI(i)					lamda diagonal		
-2	1	1	-1	0		0	1		-1	0	
-3	1	2	-0.5	0.866025		0	0		-0.5	0.866025	
0	0	-1	-0.5	-0.86603		1	0		-0.5	-0.86603	
1	2	2 values of op for irow = 1									
3	1	1 values of pass for irow = 1									
eigenvectors (real)			eigenvectors (imag)			eigenvector matrix inverse (real)			eigenvector mat inv(imag)		
0	0.5	0.5	0	-0.2887	0.2887	0	0	1	0	0	0
-1	1	1	0	0	0	0	0.5	0.5	1.7321	-0.866	-0.866
1	0	0	0	0	0	0	0.5	0.5	-1.7321	0.866	0.866
lamda matrix (real)			lamda matrix (imag)			vecTinv * vecT (real)			vecTinv * vecT (imag)		
-1	0	0	0	0	0	1	0	0	0	0	0
0	-0.5	0	0	0.866	0	0	1	0	0	0	0
0	0	-0.5	0	0	-0.866	0	0	1	0	0	0
Ac	Bc	WR	WI								
0	0	-1	0								
-0.5	0.2887	-0.5	0.866								
-0.5	-0.2887	-0.5	-0.866								

The step-response equation is: $Y = \sum_{i=1}^{3} \left\{ e^{WR_i t} [A_{C_i} \cos(WI_i t) - B_{C_i} \sin(WI_i t)] - A_{C_i} \right\}$.

Plugging numbers:
$$Y = e^{-0.5t}[-0.5\cos(0.866t) - 0.2887\sin(0.866t)] + 0.5$$
$$+ e^{-0.5t}[-0.5\cos(-0.866t) + 0.2887\sin(-0.866t)] + 0.5$$

The following is a plot of Y and its comparison with Runge-Kutta integration of the system equations.

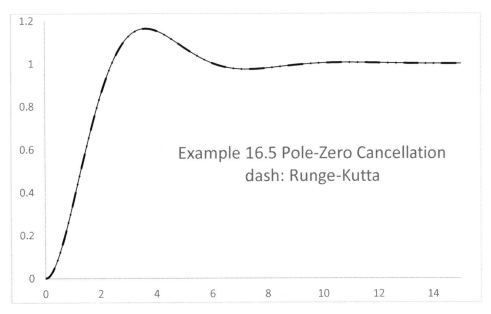

It verifies the eigenvectors. The eigenvalue at $\lambda = -1$ does not appear in the equation for Y. The reason is seen in the following Laplace transfer function from U to Y.

$$\frac{Y}{U} = \frac{(s+1)}{(s+1)(s+0.5+j0.866)(s+0.5-j0.866)}$$

The eigenvalue at $\lambda = -1$ is canceled by a zero. This means that the eigenvalue $\lambda = -1$ doesn't affect the output due to U. Eigenvectors are able to show this.

Chapter 17: Factoring a Polynomial Using Eigenvalues

The definition for eigenvalues is: $|A - \lambda I| = 0$. This is a polynomial equation that is often difficult, if not impossible, to solve in closed form. Ogata shows ways to turn a polynomial into a matrix, which can then be solved using the *Eigenvalue Problem*.[14] This is one of those ways.

Given the polynomial: $s^6 + 18s^5 + 150.75s^4 + 499.5s^3 + 303.75s^2 - 1215s - 1458 = 0$.

Multiply it by a variable: $s^6 x + 18s^5 x + 150.75s^4 x + 499.5s^3 x + 303.75s^2 x - 1215s^1 x - 1458x = 0$.

Let's let: $y_1 = x$. Then let's let:

$$sy_1 = sx = y_2$$
$$sy_2 = s^2 x = y_3$$
$$sy_3 = s^3 x = y_4$$
$$sy_4 = s^4 x = y_5$$
$$sy_5 = s^5 x = y_6$$
$$sy_6 = s^6 x = 1458y_1 + 1215y_2 - 303.75y_3 - 499.5y_4 - 150.75y_5 - 18y_6$$

We now have the following matrix:
$$\begin{bmatrix} sy_1 \\ sy_2 \\ sy_3 \\ sy_4 \\ sy_5 \\ sy_6 \end{bmatrix} = \begin{bmatrix} 0 & 1 & 0 & 0 & 0 & 0 \\ 0 & 0 & 1 & 0 & 0 & 0 \\ 0 & 0 & 0 & 1 & 0 & 0 \\ 0 & 0 & 0 & 0 & 1 & 0 \\ 0 & 0 & 0 & 0 & 0 & 1 \\ 1458 & 1215 & -303.75 & -499.5 & -150.75 & -18 \end{bmatrix} \begin{bmatrix} y_1 \\ y_2 \\ y_3 \\ y_4 \\ y_5 \\ y_6 \end{bmatrix}$$

We put it into program Sub eig_main.

The Spreadsheet for Program Sub eig_main							
6						WR(i)	WI(i)
0	1	0	0	0	0	-1.5	0
0	0	1	0	0	0	-2.998	0
0	0	0	1	0	0	-3.002	0
0	0	0	0	1	0	1.5	0
0	0	0	0	0	1	-6	6
1458	1215	-303.75	-499.5	-150.75	-18	-6	-6
eigsum	trace	prod	deter				
-18	-18	-1458	-1458				

The roots are: -1.5, -3, -3, 1.5, -6 + j6, -6 - j6.

---

[14] K. Ogata, *State Space Analysis of Control Systems* (Englewood Cliffs, NJ: Prentice Hall, 1967).

Appendix A: Deriving the Equation for Step Response Using Eigenvectors

Linear differential equations with constant coefficients can be transformed into first-order equations.[15] This appendix shows how these equations can be further transformed into uncoupled equations using eigenvectors. The step response of these uncoupled equations is easily derived. We will demonstrate using a second-order system. The method is easily expanded to nth order.

Given a system of coupled first-order differential equations:

$$\begin{bmatrix} \dot{z}_1 \\ \dot{z}_2 \end{bmatrix} = A \begin{bmatrix} z_1 \\ z_2 \end{bmatrix} + Bu \text{ where u is the input and Y = CZ is the output.}$$

The A, B, and C matrices are constant.

In matrix form, the *Eigenvalue Problem* is: AV = VL, where A is the matrix from the given system, L is the diagonal matrix containing the eigenvalues of A, and V is the matrix of corresponding eigenvectors.

Using V as a transformation matrix: $\begin{bmatrix} z_1 \\ z_2 \end{bmatrix} = V \begin{bmatrix} q_1 \\ q_2 \end{bmatrix}$.

The system equations can be written: $\begin{bmatrix} \dot{z}_1 \\ \dot{z}_2 \end{bmatrix} = V \begin{bmatrix} \dot{q}_1 \\ \dot{q}_2 \end{bmatrix} = AV \begin{bmatrix} q_1 \\ q_2 \end{bmatrix} + Bu$.

Continuing: $\begin{bmatrix} \dot{q}_1 \\ \dot{q}_2 \end{bmatrix} = V^{-1}AV \begin{bmatrix} q_1 \\ q_2 \end{bmatrix} + V^{-1}Bu = L \begin{bmatrix} q_1 \\ q_2 \end{bmatrix} + \begin{bmatrix} vB_1 \\ vB_2 \end{bmatrix} u$.

In equation form: $\dot{q}_1 = \lambda_1 q_1 + vB_1 u$ and $\dot{q}_2 = \lambda_2 q_2 + vB_2 u$.

Let's temporarily drop the subscripts and solve one of these equations with u = 1 for a unit step.

$$\dot{q} = \lambda q + vB, \text{ which can be written } \dot{q} - \lambda q = vB.$$

Multiply the last equation by the variable p.

$$p\dot{q} - \lambda pq = p*vB. \qquad \text{Equation 1}$$

Now:
$$\frac{d}{dt}(pq) = p\dot{q} + \dot{p}q. \qquad \text{Equation 2}$$

The left-hand side of **Equation 1** will equal the right-hand side of **Equation 2** if $\dot{p} = -\lambda p$. Solving this for p: $\int \frac{\dot{p}}{p} dt = \ln(p) = -\lambda t$, which means that $p = e^{-\lambda t}$.

---

[15] Ibid.

Appendix A

The left-hand side of Equation 1 has been turned into: $\dfrac{d}{dt}(pq) = \dfrac{d}{dt}(e^{-\lambda t}q)$.

Equation 1 becomes: $\dfrac{d}{dt}(e^{-\lambda t}q) = p*vB = vB*e^{-\lambda t}$.

Continuing: $q = e^{\lambda t}\displaystyle\int_0^t (vB*e^{-\lambda t})\, dt$.

Since vB is a constant: $q = vB*e^{\lambda t}\displaystyle\int_0^t e^{-\lambda t}\, dt = \dfrac{vB}{-\lambda}*e^{\lambda t}\displaystyle\int_0^t e^{-\lambda t}(-\lambda\, dt)$.

Finally: $q = \dfrac{vB}{\lambda}*(e^{\lambda t} - 1)$.   Equation 3

Reintroducing the subscripts, the system equations have now been solved.

$$\begin{bmatrix} Z_1 \\ Z_2 \end{bmatrix} = V\begin{bmatrix} q_1 \\ q_2 \end{bmatrix} = V\begin{bmatrix} \dfrac{vB_1}{\lambda_1}*(e^{\lambda_1 t} - 1) \\ \dfrac{vB_2}{\lambda_2}*(e^{\lambda_2 t} - 1) \end{bmatrix}.$$

The output can now be written:

$$Y = \begin{bmatrix} C_1 & C_2 \end{bmatrix}\begin{bmatrix} Z_1 \\ Z_2 \end{bmatrix} = \begin{bmatrix} C_1 & C_2 \end{bmatrix}V\begin{bmatrix} \dfrac{vB_1}{\lambda_1}*(e^{\lambda_1 t} - 1) \\ \dfrac{vB_2}{\lambda_2}*(e^{\lambda_2 t} - 1) \end{bmatrix} = \begin{bmatrix} Cv_1 & Cv_2 \end{bmatrix}\begin{bmatrix} \dfrac{vB_1}{\lambda_1}*(e^{\lambda_1 t} - 1) \\ \dfrac{vB_2}{\lambda_2}*(e^{\lambda_2 t} - 1) \end{bmatrix}.$$

Continuing: $Y = \dfrac{Cv_1 vB_1}{\lambda_1}*(e^{\lambda_1 t} - 1) + \dfrac{Cv_2 vB_2}{\lambda_2}*(e^{\lambda_2 t} - 1)$.

We can expand this to the nth order: $Y = \displaystyle\sum_{i=1}^{n}\left[\dfrac{Cv_i vB_i}{\lambda_i}*(e^{\lambda_i t} - 1)\right]$.   Equation 4

The eigenvector matrix V is used to compute the coefficients $\dfrac{Cv_i vB_i}{\lambda_i}$. This matrix clearly shows the contribution to the output due to each eigenvalue.

Appendix B: Deriving the Equation for Step Response with Complex Eigenvalues

Complex eigenvalues come in conjugate pairs. Hence, we need two terms from **Equation 4** in appendix A.

$$Y = \frac{Cv_1 vB_1}{\lambda_1} * (e^{\lambda_1 t} - 1) + \frac{Cv_2 vB_2}{\lambda_2} * (e^{\lambda_2 t} - 1).$$

If $\lambda_1 = R_1 + jQ_1$, $\lambda_2 = R_2 + jQ_2$, $\dfrac{Cv_1 vB_1}{\lambda_1} = a_{c1} + jb_{c1}$, and $\dfrac{Cv_2 vB_2}{\lambda_2} = a_{c2} + jb_{c2}$, then Y becomes:

$$Y = (a_{c1} + jb_{c1}) * (e^{(R_1 + jQ_1)t} - 1) + (a_{c2} + jb_{c2}) * (e^{(R_2 + jQ_2)t} - 1).$$

Substitute: $e^{(R_1 + jQ_1)t} = e^{R_1 t}[\cos(Q_1 t) + j\sin(Q_1 t)]$ and $e^{(R_2 + jQ_2)t} = e^{R_2 t}[\cos(Q_2 t) + j\sin(Q_2 t)]$ into Y and gather terms:

$$Y = \{a_{c1} e^{R_1 t} \cos(Q_1 t) - a_{c1} - b_{c1} e^{R_1 t} \sin(Q_1 t) + a_{c2} e^{R_2 t} \cos(Q_2 t) - a_{c2} - b_{c2} e^{R_2 t} \sin(Q_2 t)\}$$
$$+ j\{a_{c1} e^{R_1 t} \sin(Q_1 t) + b_{c1} e^{R_1 t} \cos(Q_1 t) - b_{c1} + a_{c2} e^{R_2 t} \sin(Q_2 t) + b_{c2} e^{R_2 t} \cos(Q_2 t) - b_{c2}\} \quad \text{Equation 1}$$

The eigenvalues are in conjugate pairs: $R_2 = R_1$ and $Q_2 = -Q_1$. The coefficients in **Equation 1** must also be in conjugate pairs: $a_{c2} = a_{c1}$ and $b_{c2} = -b_{c1}$. The complex terms vanish.

Therefore: $Y = a_{c1} e^{R_1 t} \cos(Q_1 t) - a_{c1} - b_{c1} e^{R_1 t} \sin(Q_1 t) + a_{c1} e^{R_1 t} \cos(-Q_1 t) - a_{c1} + b_{c1} e^{R_1 t} \sin(-Q_1 t).$

Continuing:
$$Y = 2\{e^{R_1 t}[a_{c1} \cos(Q_1 t) - b_{c1} \sin(Q_1 t)] - a_{c1}\}.$$

If we use the notation $Q_1 = \lambda_1$ and $R_1 = \lambda_R$ and drop the subscripts:

$$Y = 2\{e^{\lambda_R t}[a_c \cos(\lambda_1 t) - b_c \sin(\lambda_1 t)] - a_c\}. \quad \text{Equation 2}$$

Appendix C: The Equation for Combining Real and Complex Eigenvalues

Repeating the real part of Equation 1 in appendix B:

$$Y = a_{C1}e^{R_1 t}\cos(Q_1 t) - a_{C1} - b_{C1}e^{R_1 t}\sin(Q_1 t) + a_{C2}e^{R_2 t}\cos(Q_2 t) - a_{C2} - b_{C2}e^{R_2 t}\sin(Q_2 t).$$

This can be put in a summation:

$$Y = \sum_{i=1}^{2}\left\{e^{R_i t}\left[a_{Ci}\cos(Q_i t) - b_{Ci}\sin(Q_i t)\right] - a_{Ci}\right\}.$$

Now a real eigenvalue can also be put in this summation if $b_{Ci} = 0$ and $Q_i = 0$. Therefore, real and complex eigenvalues can all be combined into a single summation.

$$Y = \sum_{i=1}^{n}\left\{e^{R_i t}\left[a_{Ci}\cos(Q_i t) - b_{Ci}\sin(Q_i t)\right] - a_{Ci}\right\}.$$

When this formula is programmed, R_i is called WR_i, and Q_i is called WI_i.

$$Y = \sum_{i=1}^{n}\left\{e^{WR_i t}\left[a_{Ci}\cos(WI_i t) - b_{Ci}\sin(WI_i t)\right] - a_{Ci}\right\}. \qquad \text{Equation 1}$$

Appendix D: Source Code Listings

D.1: **Program LR** that implements the LR algorithm and is discussed and used in chapter 2.

D.2: **Program QR** that implements the QR algorithm and is discussed and used in chapter 2.

D.3: **Program eig**, which is a program that computes all of the eigenvalues of a general real matrix. It is called by **Program eig_main**, which is discussed and listed in chapter 14.

D.4: **Program eigvec**, which is a program that computes all the eigenvectors of a general real matrix. It is called by **Program eigvec_main**, which is discussed and listed in chapter 14.

Appendix D.1: A Program That Implements the LR Algorithm

Code	Comments
| ```
Option Base 1
Sub LR_main()
Dim A(), L(), R()
N = Cells(1, 1)
ReDim A(N, N), L(N, N), R(N, N)
For i = 1 To N
   For j = 1 To N
      A(i, j) = Cells(i + 1, j)
   Next j
Next i
For iter = 0 To 13

  Cells(iter * (N + 1) + 1 + 1, 3 * N + 3) = iter
    For i = 1 To N
     For j = 1 To N
       Cells(iter * (N + 1) + 1 + i, j) = Application.Round(A(i, j), 2)
     Next j
   Next i
  Call L_R(N, A, L, R)
    For i = 1 To N
     For j = 1 To N
       Cells(iter * (N + 1) + 1 + i, j + 1 * N + 1) = Application.Round(L(i, j), 2)
       Cells(iter * (N + 1) + 1 + i, j + 2 * N + 2) = Application.Round(R(i, j), 2)
     Next j
   Next i
  A = Application.MMult(R, L)

Next iter
End Sub
``` | • Read in N and A(I, j).<br><br>• For iter = 0 To 13<br><br><br><br><br><br><br>• Call sub L_R(N, A, L, R)<br><br><br><br><br><br>• A ← R ∗ L<br><br>• Next iter |
| ```
Sub L_R(N, A, L, R)
  For i = 1 To N
    For j = 1 To N
      L(i, j) = 0: R(i, j) = 0
    Next j
    L(i, i) = 1
  Next i
  For k = 1 To N - 1
      If Abs(A(k, k)) < 1e-06 Then A(k, k) = Sgn(A(k, k)) * 0.001
    For i = k + 1 To N
      L(i, k) = A(i, k) / A(k, k)
      For j = k + 1 To N
         A(i, j) = A(i, j) - L(i, k) * A(k, j)
      Next j
    Next i
  Next k
  For i = 1 To N
    For j = i To N
      R(i, j) = A(i, j)
    Next j
  Next i
End Sub
``` | • Subprogram L_R factors **A** into:<br><br>• **L**, a lower triangular matrix<br><br>• **R**, an upper triangular matrix<br><br>• See the following reference for LR factorization. Mathews, J., and K. Fink. *Numerical Methods Using MATLAB*. Prentice-Hall, 1999.<br><br>• Note the *poor man's work-around* for *division by zero*.<br><br>• End Sub |

Appendix D.2: A Program That Implements the QR Algorithm

Subprogram QR_main
- Read in N and A(I, j)
- For iter = 0 To (Niter = 11)
  - call Sub Q_R(N, A, Q, R)
  - A ← R * Q
- Next iter

End Sub

Subprogram Q_R factors A into:

- **Q**, a Householder Orthogonal matrix
- **R**, an upper triangular matrix
- See the following reference for QR factorization: Hager, W., *Applied Numerical Linear Algebra*. Prentice-Hall, 1988.

End Subprogram

```
Option Base 1
Sub QR_main()
Dim A(), Q(), R()
N = Cells(1, 1)
ReDim A(N, N), Q(N, N), R(N, N)
For i = 1 To N
  For j = 1 To N
    A(i, j) = Cells(i + 1, j)
  Next j
Next i
For iter = 0 To 11
  Cells(iter * (N + 1) + 1 + 1, 3 * N + 3) = iter
  For i = 1 To N
    For j = 1 To N
      Cells(iter * (N + 1) + 1 + i, j) = Application.Round(A(i, j), 2)
    Next j
  Next i
  Call Q_R(N, A, Q, R)
  A = Application.MMult(R, Q)
Next iter
End Sub ' QR_main
```

```
Sub Q_R(N, A, QM, RM)
Dim d(20), AQ(20, 20), Ident(20, 20), vh(), vhT(), H()
ReDim vh(N, 1), vhT(1, N), H(N, N)
 For i = 1 To N
   For j = 1 To N
     QM(i, j) = 0: RM(i, j) = 0: AQ(i, j) = 0: Ident(i, j) = 0
   Next j
   QM(i, i) = 1: Ident(i, i) = 1
 Next i
 k = 0
 For L = 1 To N              """ Begin Transform
  k = k + 1
  If k = N Then
    d(L) = A(k, L): Exit For
  End If
  sarg = 0
  For i = k To N
    sarg = sarg + A(i, L) ^ 2
  Next i
  s = Sqr(sarg)
  If s = 0 Then
    d(L) = 0: GoTo nextL
  End If
  T = A(k, L): R = 1 / Sqr(s * (s + Abs(T)))
  If T < 0 Then s = -s
  d(L) = -s: A(k, k) = R * (T + s)
  For i = k + 1 To N
    A(i, k) = R * A(i, L)
  Next i
  For j = L + 1 To N
    T = 0
    For i = k To N
      T = T + A(i, k) * A(i, j)
    Next i
    For i = k To N
      A(i, j) = A(i, j) - T * A(i, k)
    Next i
  Next j
nextL: Next L            """ End Transform

For i = 1 To N
  RM(i, i) = d(i)
Next i
For i = 1 To N - 1
  For j = i + 1 To N
    RM(i, j) = A(i, j)
  Next j
Next i
For j = 1 To N
  For i = j To N
    AQ(i, j) = A(i, j)
  Next i
Next j
For j = 1 To N - 1
  For i = 1 To N
    vh(i, 1) = AQ(i, j): vhT(1, i) = vh(i, 1)
  Next i
  vhvhT = Application.MMult(vh, vhT)
  For ii = 1 To N
    For jj = 1 To N
      H(ii, jj) = Ident(ii, jj) - vhvhT(ii, jj)
    Next jj
  Next ii
  Qtot = Application.MMult(QM, H)
  For ii = 1 To N
    For jj = 1 To N
      QM(ii, jj) = Qtot(ii, jj)
    Next jj
  Next ii
Next j
End Sub ' Q_R
```

Appendix D.3: Sub eig, Page 1 of 6	Sub eig, Page 2 of 6
| ```
Sub eig(npr, Ainput, WR, WI)
Dim Q(), R(), Ahess(), Shift(), Asave(100, 100), A()
N = npr: itype = 0: iroot=0: tol = 1e-06: niter = 500
ReDim Q(N, N), R(N, N), Ahess(N, N), Shift(N, N), A(N, N)
For i = 1 To N
  For j = 1 To N
    A(i, j) = Ainput(i, j)
    Ahess(i, j) = A(i, j)
  Next j
Next i
If N < 3 Then GoTo BOTTOM
Call hess(N, Ahess)
For i = 1 To N
  For j = 1 To N
    A(i, j) = Ahess(i, j)
  Next j
Next i
For iter = 1 To niter          '''' BEGIN QR ITERATIONS
 If iter = 100 Then tol = 0.00001
 If iter = 200 Then tol = 0.001
 If iter = 300 Then tol = 0.01
   If iter < 100 Then               " shift begin
     aR = A(N - 1, N - 1): bR = A(N - 1, N)
     cR = A(N, N - 1): dR = A(N, N)
   Else
     aR = A(N - 1, N - 2): bR = A(N - 1, N - 1)
     cR = A(N, N - 2): dR = A(N, N - 1)
   End If
   RAD = dR ^ 2 + aR ^ 2 - 2 * aR * dR + 4 * bR * cR
   If RAD >= 0 Then              '''' two REALs
     WR1 = 0.5 * (dR + aR + Sqr(RAD)): d1 = Abs(WR1 - A(N, N))
     WR2 = 0.5 * (dR + aR - Sqr(RAD)): d2 = Abs(WR2 - A(N, N))
     If d1 < d2 Then
       guess = WR1
     Else
       guess = WR2
     End If
   Else                          '''' one COMPLEX
     WR1 = 0.5 * (dR + aR): WI1 = 0.5 * Sqr(-RAD)
     guess = Sqr(WR1 ^ 2 + WR2 ^ 2)
   End If
   For i = 1 To N
     For j = 1 To N
       Shift(i, j) = 0
     Next j
     Shift(i, i) = guess
   Next i
   For i = 1 To N
     For j = 1 To N
       A(i, j) = A(i, j) - Shift(i, j)
     Next j
   Next i                            " shift end
 Call Q_R(N, A, Q, R)
   A = Application.MMult(R, Q)
   For i = 1 To N                '''' unshift begin
     For j = 1 To N
       A(i, j) = A(i, j) + Shift(i, j)
     Next j
   Next i                         '''' unshift end
 Call deflate(N, A, itype, iroot, WR, WI, tol)
``` | ```
 If itype <> 0 Then           '''' Begin "A" Deflate
   If itype = 1 Then
     For i = 1 To N - 1
       For j = 1 To N - 1
         Asave(i, j) = A(i, j)
       Next j
     Next i
     N = N - 1
   Else                              ''' itype = 2
     For i = 1 To N - 2
       For j = 1 To N - 2
         Asave(i, j) = A(i, j)
       Next j
     Next i
     N = N - 2
   End If
   ReDim A(N, N), Q(N, N), R(N, N)
   For i = 1 To N
     For j = 1 To N
       A(i, j) = Asave(i, j)
     Next j
   Next i
   itype = 0
 End If                        ''''    End "A" Deflate
 If N <= 2 Then Exit For        ''''        SUCCESS
Next iter                       ''''   End QR Iterations

If N > 2 Then    ' here iroot <> npr, so sub eig has failed
 Cells(1, npr + 5) = " Sub eig": Cells(1, npr + 6) = " fails"
 Cells(2, npr + 5) = " iter =": Cells(2, npr + 6) = iter
 Cells(3, npr + 5) = " iroot =": Cells(3, npr + 6) = iroot
 Cells(4, npr + 5) = " WR =": Cells(4, npr + 6) = " WI ="
 For i = 1 To iroot
   Cells(4 + i, npr + 5) = WR(i): Cells(4 + i, npr + 6) = WI(i)
 Next i
 Cells(npr + 4, npr + 2) = iter
 For i = 1 To N
   For j = 1 To N
     Cells(npr + 3 + i, j) = A(i, j)
   Next j
 Next i
 End                              '''' Sub eig fails
End If

BOTTOM:   '        compute the last 1 or 2 eigenvalues
 If N = 1 Then                    ''''   one REAL
   iroot = iroot + 1: WR(iroot) = A(N, N): WI(iroot) = 0
 Else                             ' N=2 two roots
   aR = A(N - 1, N - 1): bR = A(N - 1, N)
   cR = A(N, N - 1): dR = A(N, N)
   RAD = dR ^ 2 + aR ^ 2 - 2 * aR * dR + 4 * bR * cR
   If RAD >= 0 Then                '''' two REALs
     WR1 = 0.5 * (dR + aR + Sqr(RAD)): WI1 = 0
     WR2 = 0.5 * (dR + aR - Sqr(RAD)): WI2 = 0
   Else                            '''' one COMPLEX
     WR1 = 0.5 * (dR + aR): WI1 = 0.5 * Sqr(-RAD)
     WR2 = WR1: WI2 = -WI1
   End If
   iroot = iroot + 1: WR(iroot) = WR1: WI(iroot) = WI1
   iroot = iroot + 1: WR(iroot) = WR2: WI(iroot) = WI2
 End If                            '  N= 1 or 2
End Sub ' eig
``` |

Appendix D.3: Sub eig, Page 3 of 6

```
Sub hess(N, AA)
  LA = N - 1: eps = 2 ^ (-52): T = 0
  For m = 2 To LA
    i = m: x = 0
    For j = m To N
      If Abs(AA(j, m - 1)) > Abs(x) Then
        x = AA(j, m - 1): i = j
      End If
    Next j
    If i <> m Then
      For j = (m - 1) To N
        y = AA(i, j)
        AA(i, j) = AA(m, j)
        AA(m, j) = y
      Next j
      For j = 1 To N
        y = AA(j, i)
        AA(j, i) = AA(j, m)
        AA(j, m) = y
      Next j
    End If
    If x <> 0 Then
      For i = (m + 1) To N
        y = AA(i, m - 1)
        If y <> 0 Then
          AA(i, m - 1) = y / x
          y = AA(i, m - 1)
          For j = m To N
            AA(i, j) = AA(i, j) - y * AA(m, j)
          Next j
          For j = 1 To N
            AA(j, m) = AA(j, m) + y * AA(j, i)
          Next j
        End If
      Next i
    End If
  Next m
  For jj = 1 To N - 2
    For ii = jj + 2 To N
      AA(ii, jj) = 0
    Next ii
  Next jj
End Sub ' hess
```

Sub eig, Page 4 of 6

```
Sub deflate(N, A, itype, iroot, WR, WI, tol)
  ctr = 0
  For j = 1 To N - 1
    If j < N - 1 Then
      If Abs(A(N - 1, j)) < tol Then ctr = ctr + 1 ' on each pass j < N - 1
      If Abs(A(N, j)) < tol Then GoTo NEXTJ
      Exit For                           ' return to QR W/O success
    End If
                                         '''''' here j = N - 1
    If Abs(A(N, j)) < tol Then           ' begin j = N - 1 checks
      iroot = iroot + 1: WR(iroot) = A(N, N): WI(iroot) = 0 '''''' one REAL
      itype = 1: Exit For                ' return to QR to deflate
    End If
    If ctr = N - 2 Then                  '''''' two roots
      aR = A(N - 1, N - 1): bR = A(N - 1, N)
      cR = A(N, N - 1): dR = A(N, N)
      RAD = dR ^ 2 + aR ^ 2 - 2 * aR * dR + 4 * bR * cR
      If RAD >= 0 Then                   '''''' two REALs
        WR1 = 0.5 * (dR + aR + Sqr(RAD)): WI1 = 0
        WR2 = 0.5 * (dR + aR - Sqr(RAD)): WI2 = 0
      Else                               '''''' one COMPLEX
        WR1 = 0.5 * (dR + aR): WI1 = 0.5 * Sqr(-RAD)
        WR2 = WR1: WI2 = -WI1
      End If
      iroot = iroot + 1: WR(iroot) = WR1: WI(iroot) = WI1
      iroot = iroot + 1: WR(iroot) = WR2: WI(iroot) = WI2
      itype = 2: Exit For                ' return to QR to deflate
    End If                               ' ctr=N-2
    Exit For                             ''' return to QR W/O success
NEXTJ:
  Next j
End Sub ' deflate
```

Appendix D.3: Sub eig, Page 5 of 6	Sub eig, Page 6 of 6
| ```
Sub Q_R(N, A, QM, RM)
Dim d(100), AQ(100, 100), Ident(100, 100), vh(), vhT(), H()
ReDim vh(N, 1), vhT(1, N), H(N, N)
 For i = 1 To N
   For j = 1 To N
     QM(i, j) = 0: RM(i, j) = 0: AQ(i, j) = 0: Ident(i, j) = 0
   Next j
     QM(i, i) = 1: Ident(i, i) = 1
 Next i
 k = 0
 For L = 1 To N          '''' Begin Transform
   k = k + 1
   If k = N Then
     d(L) = A(k, L): Exit For
   End If
   sarg = 0
   For i = k To N
     sarg = sarg + A(i, L) ^ 2
   Next i
   s = Sqr(sarg)
   If s = 0 Then
     d(L) = 0: GoTo nextL
   End If
   T = A(k, L): R = 1 / Sqr(s * (s + Abs(T)))
   If T < 0 Then s = -s
   d(L) = -s: A(k, k) = R * (T + s)
   For i = k + 1 To N
    A(i, k) = R * A(i, L)
   Next i
   For j = L + 1 To N
     T = 0
     For i = k To N
      T = T + A(i, k) * A(i, j)
     Next i
     For i = k To N
       A(i, j) = A(i, j) - T * A(i, k)
     Next i
   Next j
nextL: Next L           '''' End Transform
``` | ```
 For i = 1 To N
   RM(i, i) = d(i)
 Next i
 For i = 1 To N - 1
   For j = i + 1 To N
     RM(i, j) = A(i, j)
   Next j
 Next i
 For j = 1 To N
   For i = j To N
     AQ(i, j) = A(i, j)
   Next i
 Next j
 For j = 1 To N - 1
   For i = 1 To N
     vh(i, 1) = AQ(i, j): vhT(1, i) = vh(i, 1)
   Next i
   vhvhT = Application.MMult(vh, vhT)
   For ii = 1 To N
     For jj = 1 To N
       H(ii, jj) = Ident(ii, jj) - vhvhT(ii, jj)
     Next jj
   Next ii
   Qtot = Application.MMult(QM, H)
   For ii = 1 To N
     For jj = 1 To N
       QM(ii, jj) = Qtot(ii, jj)
     Next jj
   Next ii
 Next j
End Sub ' QR
``` |

Appendix D.4: Sub eigvec, page 1 of 7	Sub eigvec, page 2 of 7
| ```
Sub eigvec(K2, areal, lamda)
    Dim A(20, 20), Ain(20, 20), ainsave(20, 20)
    Dim vec(20), ax(20), row(20), ad(20, 20), bd(20), xd(20)
    Dim vecT(20, 20), axT(20, 20), vecTinv(20, 20), Lout(20, 20)
    Dim lout(20, 20), xL(20), xLT(20, 20)
    N = K2 - 1: minus = Application.Complex(-1, 0)
    irow = 1
Vtop:
For lam = 1 To K2
    For i = 1 To K2
        For j = 1 To K2
            Ain(i, j) = areal(i, j)
        Next j
    Next i
    For i = 1 To K2
        Ain(i, i) = Application.ImSub(areal(i, i), lamda(lam))
    Next i
    For i = 1 To K2
     For j = 1 To K2
         ainsave(i, j) = Ain(i, j)
      Next j
    Next i
    op = 1
ontop:
    pass = 1
top:
    Call crow(row, K2, N, pass, irow)
    For i = 1 To N
        For j = 1 To K2
           A(i, j) = Ain(row(i), j)
        Next j
    Next i
    For i = 1 To N
        For j = 1 To N
           ad(i, j) = A(i, j)
        Next j
         bd(i) = Application.ImProduct(minus, A(i, K2))
    Next i

    Call lineq(ad, bd, xd, N, fail)
``` | ```
If fail = 0 Then
    GoTo Bot
Else
    pass = pass + 1
    If pass > K2 Then
        op = op + 1
        If op > K2 Then
            Cells(K2 + 2, 1) = "Eigenvector could not be computed"
            End
        End If
        For i = 1 To K2
            For j = 1 To K2
                Ain(i, j) = ainsave(i, j)
            Next j
        Next i
        For i = 1 To K2
            tempC = Ain(i, K2 - op + 1)
            Ain(i, K2 - op + 1) = Ain(i, K2)
            Ain(i, K2) = tempC
        Next i
        GoTo ontop
    End If
    GoTo top
End If
``` |

Appendix D.4: Sub eigvec, page 3 of 7

```
Bot:
  If irow = 1 Then
    Cells(K2 + 4, lam) = op
    Cells(K2 + 5, lam) = pass
  Else
    Cells(K2 + 7, lam) = op
    Cells(K2 + 8, lam) = pass
  End If
  For i = 1 To N
    vec(i) = xd(i)
  Next i
    vec(K2) = Application.Complex(1, 0)
  If op > 1 Then
    temp = vec(K2)
    vec(K2) = vec(K2 - op + 1)
    vec(K2 - op + 1) = temp
  End If
  Vmax = Application.ImAbs(vec(1))
  For i = 2 To K2
    If Application.ImAbs(vec(i)) > Vmax Then
      Vmax = Application.ImAbs(vec(i))
    End If
  Next i
  VmaxC = Application.Complex(Vmax, 0)
  For i = 1 To K2
    vec(i) = Application.ImDiv(vec(i), VmaxC)
  Next i

  For i = 1 To K2
    dum1 = Application.Complex(0, 0)
    For j = 1 To K2
      dum2 = Application.ImProduct(areal(i, j), vec(j))
      dum1 = Application.ImSum(dum1, dum2)
    Next j
    ax(i) = dum1
    xL(i) = Application.ImProduct(vec(i), lamda(lam))
  Next i

  For i = 1 To K2
    vecT(i, lam) = vec(i)
    axT(i, lam) = ax(i)
    xLT(i, lam) = xL(i)
    vecR = Application.ImReal(vec(i))
    vecI = Application.Imaginary(vec(i))
    Cells(i + K2 + 10, lam) = Application.Round(vecR, 4)
    Cells(i + K2 + 10, lam + K2 + 1) = Application.Round(vecI, 4)
  Next i

Next lam
```

Sub eigvec, page 4 of 7

```
Call inverse_main(vecT, vecTinv, K2, fail)
Cells(K2 + 4, K2 + 1) = "values of op for irow = 1"
Cells(K2 + 5, K2 + 1) = "values of pass for irow = 1"

If fail = 1 Then
  If irow = 1 Then
    irow = 2: GoTo Vtop
  Else
    Call noinverse(K2, axT, xLT)
  End If
End If

Call mult(vecTinv, axT, Lout, K2, K2, K2)
For i = 1 To K2
  For j = 1 To K2
    LoutR = Application.ImReal(Lout(i, j))
    LoutI = Application.Imaginary(Lout(i, j))
    Cells(i + 2 * K2 + 12, j) = Application.Round(LoutR, 4)
    Cells(i + 2 * K2 + 12, j + K2 + 1) = Application.Round(LoutI, 4)
    vinvR = Application.ImReal(vecTinv(i, j))
    vinvI = Application.Imaginary(vecTinv(i, j))
    Cells(i + K2 + 10, j + 2 * (K2 + 1)) = Application.Round(vinvR, 4)
    Cells(i + K2 + 10, j + 3 * (K2 + 1)) = Application.Round(vinvI, 4)
  Next j
  LdiagR = Application.ImReal(Lout(i, i))
  LdiagI = Application.Imaginary(Lout(i, i))
  Cells(i + 1, K2 + 9) = LdiagR
  Cells(i + 1, K2 + 10) = LdiagI
Next i

Call mult(vecTinv, vecT, Iout, K2, K2, K2)
For i = 1 To K2
  For j = 1 To K2
    IoutR = Application.ImReal(Iout(i, j))
    IoutI = Application.Imaginary(Iout(i, j))
    Cells(i + 2 * K2 + 12, j + 2 * K2 + 2) = Application.Round(IoutR, 3)
    Cells(i + 2 * K2 + 12, j + 3 * K2 + 3) = Application.Round(IoutI, 3)
  Next j
Next i
Call the_labels(K2, irow)

Bcnt = 0
For i = 1 To K2
  Breal = Cells(i + 1, K2 + 5)
  If Breal = 0 Then Bcnt = Bcnt + 1
Next i
If Bcnt = K2 Then
  End
Else
  Call coeff_step(K2, vecT, vecTinv, Lout)
End If
End Sub ' eigvec
```

Appendix D.4: Sub eigvec, page 5 of 7	Sub eigvec, page 6 of 7
| ```
Sub crow(row, K2, N, pass, irow)
If irow = 1 Then            ' top down
  If pass = 1 Then
    For i = 1 To N
      row(i) = K2 + 1 - i
    Next i
  Else
    row(K2 - pass + 1) = pass - 1
  End If
Else                        ' bottom up
  If pass = 1 Then
    For i = 1 To N
      row(i) = i
    Next i
  Else
    row(K2 + 1 - pass) = K2 - pass + 2
  End If
End If
End Sub ' crow

Sub inverse_main(vecT, vecTinv, K2, fail)
  Dim aR(20, 40), aI(20, 40)
  Dim vR(20, 20), vI(20, 20), vecA(20, 40)
  For i = 1 To K2
    For j = 1 To K2
      vR(i, j) = Application.ImReal(vecT(i, j))
      vI(i, j) = Application.Imaginary(vecT(i, j))
    Next j
  Next i
  For i = 1 To K2
    For j = 1 To K2 + K2
      aR(i, j) = 0
      aI(i, j) = 0
    Next j
  Next i
  For i = 1 To K2
    For j = 1 To K2
      aR(i, j) = vR(i, j): aI(i, j) = vI(i, j)
    Next j
  Next i
  For i = 1 To K2
    aR(i, i + K2) = 1
  Next i
  For i = 1 To K2
    For j = 1 To K2 + K2
      vecA(i, j) = Application.Complex(aR(i, i), aI(i, i))
    Next j
  Next i
  Call inverse(vecA, vecTinv, K2, fail)
End Sub ' inverse_main
``` | ```
Sub coeff_step(K2, vecT, vecTinv, Lout)
Dim Bx(20, 1), Cx(1, 20), vB(20, 1), Cv(1, 20)
For i = 1 To K2
  Breal = Cells(i + 1, K2 + 5)
  Creal = Cells(i + 1, K2 + 7)
  Bx(i, 1) = Application.Complex(Breal, 0)
  Cx(1, i) = Application.Complex(Creal, 0)
Next i
Call mult(vecTinv, Bx, vB, K2, K2, 1)
Call mult(Cx, vecT, Cv, 1, K2, K2)
For i = 1 To K2
  If Lout(i, i) <> 0 Then
    coeff = Application.ImProduct(Cv(1, i), vB(i, 1))
    coeff = Application.ImDiv(coeff, Lout(i, i))
  Else
    coeff = Application.Complex(0, 0)
  End If
  coeffR = Application.ImReal(coeff)
  coeffI = Application.Imaginary(coeff)
  Cells(i + 3 * K2 + 14, 1) = Application.Round(coeffR, 4)
  Cells(i + 3 * K2 + 14, 2) = Application.Round(coeffI, 4)
  LdiagR = Application.ImReal(Lout(i, i))
  LdiagI = Application.Imaginary(Lout(i, i))
  Cells(i + 3 * K2 + 14, 3) = Application.Round(LdiagR, 4)
  Cells(i + 3 * K2 + 14, 4) = Application.Round(LdiagI, 4)
Next i
Cells(3 * K2 + 14, 1) = "    Ac": Cells(3 * K2 + 14, 2) = "    Bc"
Cells(3 * K2 + 14, 3) = "    WR": Cells(3 * K2 + 14, 4) = "    WI"
End Sub ' coeff_step


Sub lineq(ad, bd, x, N, fail)
  See chapter 10 for the listing.
End Sub


Sub inverse(A, ainv, N, fail)
  See chapter 10 for the listing.
End Sub


Sub mult(A, b, ab, m, L, N)
  See chapter 10 for the listing.
End Sub
``` |

Appendix D.4: Sub eigvec, page 7 of 7

```
Sub noinverse(K2, axT, xLT)
    For i = 1 To K2
        For j = 1 To K2
            axR = Application.ImReal(axT(i, j))
            axI = Application.Imaginary(axT(i, j))
            Cells(i + 2 * K2 + 12, j) = Application.Round(axR, 4)
            Cells(i + 2 * K2 + 12, j + K2 + 1) = Application.Round(axI, 4)
            xLR = Application.ImReal(xLT(i, j))
            xLI = Application.Imaginary(xLT(i, j))
            Cells(i + 2 * K2 + 12, j + 2 * K2 + 2) = Application.Round(xLR, 4)
            Cells(i + 2 * K2 + 12, j + 3 * K2 + 3) = Application.Round(xLI, 4)
        Next j
    Next i
    Cells(K2 + 7, K2 + 1) = "values of op for irow = 2. Hence, there is no vector inverse"
    Cells(K2 + 8, K2 + 1) = "values of pass for irow = 2"
    Cells(K2 + 10, 1) = "eigenvectors (real)"
    Cells(K2 + 10, K2 + 2) = "eigenvalues (imag)"
    Cells(2 * K2 + 12, 1) = "A*V (real)"
    Cells(2 * K2 + 12, 1 + K2 + 1) = "A*V (imag)"
    Cells(2 * K2 + 12, 1 + 2 * K2 + 2) = "V*L (real)"
    Cells(2 * K2 + 12, 1 + 3 * K2 + 3) = "V*L (imag)"
    End
End Sub ' noinverse

Sub the_labels(K2, irow)
    Cells(K2 + 12, 1) = "eigenvectors (real)"
    Cells(K2 + 12, K2 + 2) = "eigenvectors (imag)"
    Cells(2 * K2 + 14, 1) = "lamda matrix (real)"
    Cells(2 * K2 + 14, K2 + 2) = "lamda matrix (imag)"
    Cells(K2 + 12, 2 * K2 + 3) = "eigenvector matrix inverse (real)"
    Cells(K2 + 12, 3 * K2 + 4) = "eigenvector mat inv(imag)"
    Cells(2 * K2 + 14, 2 * K2 + 3) = "vecTinv * vecT (real)"
    Cells(2 * K2 + 14, 3 * K2 + 4) = "vecTinv * vecT (imag)"
    Cells(1, K2 + 9) = "lamda diagonal"
    If irow = 2 Then
        Cells(K2 + 9, K2 + 1) = "values of op for irow=2"
        Cells(K2 + 10, K2 + 1) = "values of pass for irow=2"
    End If
End Sub ' the_labels
```

Appendix E

Appendix E: Code Fragments (data...Return and system...Return) for Program GoSub_main in chapter 11.
This program implements Runge-Kutta Numerical Integration.

Chapter 15 Case 1	Chapter 15 Case 2

data:
  N = 4: y(1) = 0 / 57.296: y(2) = 0 / 57.296
  y(3) = 0 / 57.296: y(4) = 0 / 57.296
  runm = Array(1, 0.5, 0.5, 1): jint = 1
  dt = 0.05: tstop = 10: istop = Int(tstop / dt)
  timeh = 0: timeX(1) = timeh
  atheta = 0: aU = 440: Aw = 0: aQ = 230: aM = 5900
  ax = 1955000: az = 4200000: aXZ = 0
  aG = 32.174: aSf = 2400: ab = 130
  qsb = aQ * aSf * ab: bou = 0.5 * ab / aU
  del = 1 - aXZ ^ 2 / (ax * az): xzx = aXZ / ax: xzz = aXZ / az
  Clp = -0.38: Clr = 0.086: Clb = -0.057: Cldr = 0.0131: Clda = 0.6
  Cnp = -0.0228: Cnr = -0.107: Cnb = 0.096: Cndr = -0.08
  Cnda = -0.01: Cyb = -0.6: Cydr = 0.171: Cyda = 0
  Yb = aQ * aSf * Cyb / aM
  Ydr = aQ * aSf * Cydr / aM: Yda = aQ * aSf * Cyda / aM
  Nb = qsb * Cnb / az: Ndr = qsb * Cndr / az: Nda = qsb * Cnda / az
  Lb = qsb * Clb / ax: Ldr = qsb * Cldr / ax: Lda = qsb * Clda / ax
  Np = qsb * bou * Cnp / az: Nr = qsb * bou * Cnr / az
  Lp = qsb * bou * Clp / ax: Lr = qsb * bou * Clr / ax
  Lpp = (Lp + xzx * Np) / del: Npp = (Np + xzz * Lp) / del
  Lrp = (Lr + xzx * Nr) / del: Nrp = (Nr + xzz * Lr) / del
  Lbp = (Lb + xzx * Nb) / del: Nbp = (Nb + xzz * Lb) / del
  Ldrp = (Ldr + xzx * Ndr) / (4 * del)
  Ndrp = (Ndr + xzz * Ldr) / (4 * del)
  Ldap = (Lda + xzx * Nda) / del: Ndap = (Nda + xzz * Lda) / del
  Kp = 0.1: Kr = 0.2: Kfe = 0.1: Fec = 5 / 57.296
Return
system:
  beta = y(1): RR = y(2): Fe = y(3): PP = y(4)
  da = Kfe * (Fec - Fe) - Kp * PP
  dR = Kr * RR
  pdot = Lpp * PP + Lrp * RR + Lbp * beta + Ldrp * dR + Ldap * da
  rdot = Npp * PP + Nrp * RR + Nbp * beta + Ndrp * dR + Ndap * da
  betad1 = Aw * PP - aU * RR + aG * Cos(atheta) * Fe + Yb * beta
  betad2 = Ydr * dR + Yda * da
  betad = (betad1 + betad2) / aU
  yd(1) = betad: yd(2) = rdot: yd(3) = PP: yd(4) = pdot
Return
	data:
	  z1 = 0.4: z2 = 0.4: w1 = 3: w2 = 6: z3 = 0.5: w3 = 0.4
	  b1 = 52.85: b2 = 25.49: b3 = -12.68
	  b4 = 54.63: b5 = 48.11: b6 = 3.02
	  b7 = 4.93: b8 = -0.93: b9 = 0.1
	  s5 = 2 * (z1 * w1 + z2 * w2 + z3 * w3)
	  s4 = w1 ^ 2 + w2 ^ 2 + w3 ^ 2 + 4 * z1 * z2 * w1 * w2 _
	    + 4 * z3 * w3 * (z1 * w1 + z2 * w2)
	  s3 = 2 * (z1 * w1 * w2 ^ 2 + z2 * w2 * w1 ^ 2) _
	    + 2 * z3 * w3 * (w1 ^ 2 + w2 ^ 2 + 4 * z1 * z2 * w1 * w2) _
	    + 2 * w3 ^ 2 * (z1 * w1 + z2 * w2)
	  s2 = w1 ^ 2 * w2 ^ 2 _
	    + 4 * z3 * w3 * (z1 * w1 * w2 ^ 2 + z2 * w2 * w1 ^ 2) _
	    + w3 ^ 2 * (w1 ^ 2 + w2 ^ 2 + 4 * z1 * z2 * w1 * w2)
	  s1 = 2 * z3 * w3 * w1 ^ 2 * w2 ^ 2 _
	    + 2 * w3 ^ 2 * (z1 * w1 * w2 ^ 2 + z2 * w2 * w1 ^ 2)
	  s0 = w1 ^ 2 * w2 ^ 2 * w3 ^ 2
	  e1 = b1 * b5 - b2 * b4
	  e2 = b2 * b6 + b5 * b3
	  e3 = b4 * b3 + b1 * b6
	  e4 = b5 * b7 + b1 * b7
	  e5 = b3 * b8 + b6 * b9
	  temp1 = b7 * e1 - b8 * e2 - b9 * e3
	  c5 = s0 / temp1
	  c6 = s1 / temp1
	  pmat(1, 1) = b3: pmat(1, 2) = b6
	  pmat(2, 1) = e2: pmat(2, 2) = e3
	  qmatD(1, 1) = s5 - b7 * c6: qmatD(2, 1) = -s3 - c6 * (e4 - e5)
	  qmatP(1, 1) = s4 - b7 * c5 + b1 + b5
	  qmatP(2, 1) = -s2 - c5 * (e4 - e5) + e1
	  pmatinv = Application.MInverse(pmat)
	  c2c4 = Application.MMult(pmatinv, qmatD)
	  c1c3 = Application.MMult(pmatinv, qmatP)
	  kT1d = c2c4(1, 1): kT1 = c1c3(1, 1)
	  kT2d = c2c4(2, 1): kT2 = c1c3(2, 1): kXd = c6: kX = c5
	  N = 6: y(1) = 0: y(2) = 0: y(3) = 0: y(4) = 0: y(5) = 0: y(6) = 0
	  T1 = y(1): T2 = y(2)
	  timeh = 0: dt = 0.02: tstop = 20: istop = Int(tstop / dt)
	  timeX(1) = timeh: Xt1(1) = T1 * 57.296: Xt2(1) = T2 * 57.296
	  Xt3(1) = 0: Xt4(1) = 0: Xt5(1) = 0
	  Xt6(1) = 0: Xt7(1) = 0: Xt8(1) = 0
	  runm = Array(1, 0.5, 0.5, 1): jint = 1
	Return
	system:
	  T1 = y(1): T2 = y(2): T1D = y(3): T2D = y(4): xx = y(5): xxD = y(6)
	  XCL = -kX * (xx - 0.5) - kXd * xxD
	  T1CL = -kT1 * T1 - kT1d * T1D
	  T2CL = -kT2 * T2 - kT2d * T2D
	  u = T1CL + T2CL + XCL
	  T1DD = b1 * T1 - b2 * T2 + b3 * u
	  T2DD = -b4 * T1 + b5 * T2 + b6 * u
	  xxDD = b8 * T1 + b9 * T2 + b7 * u
	  yd(1) = T1D: yd(2) = T2D: yd(3) = T1DD
	  yd(4) = T2DD: yd(5) = xxD: yd(6) = xxDD
	Return

Appendix E

Appendix E: Code Fragments (data...Return and system...Return) for Program GoSub_main in chapter 11. This program implements Runge-Kutta Numerical Integration.	
Chapter 15 Case 3	Chapter 15 Case 4

data:
  N = 14: y(1) = 0: y(2) = 0: y(3) = 0: y(4) = 0: y(5) = 0
  y(6) = 0: y(7) = 0: y(8) = 0: y(9) = 0: y(10) = 0: y(11) = 0
  y(12) = 0: y(13) = 0: y(14) = 0
  timeh = 0: dt = 0.01: tstop = 5: istop = Int(tstop / dt)
  timeX(1) = timeh: Xt1(1) = y(14): Xt2(1) = 0
  runm = Array(1, 0.5, 0.5, 1): jint = 1
  vel = 1300: thr = 420000: tc = 340000: aero = 200000
  mass = 5500
  tom = thr / mass: aom = aero / mass: com = tc / mass
  mua = 3.5: muc = 4.6: mup = 0.06: zfs = 0.01: wfs = 5.1
  wfsq = wfs ^ 2: fsd = 17.9: tzofs = 2 * zfs * wfs
  meq1 = 1590: zb1 = 0.01: wb1 = 20: sigma1 = 0.01
  comeq1 = tc / meq1: tzob1 = 2 * zb1 * wb1: wbsq1 = wb1 ^ 2
  meq2 = 800: zb2 = 0.01: wb2 = 30: sigma2 = 0.005
  comeq2 = tc / meq2: tzob2 = 2 * zb2 * wb2: wbsq2 = wb2 ^ 2
  wf = 20: Zf = 0.5: kd = 2.1: Kr = 0.51: wkd = 25: tfin = 0
  zact = 0.7: wact = 60: wf = 70: Kr = 0.7: wkd = 35
Return
 	
system:
  thetaTOT = y(1) + sigma1 * y(6) + sigma2 * y(8)
  Bonly = sigma1 * y(6)
  thetadTOT = y(2) + sigma1 * y(7) + sigma2 * y(9)
  deltac = (1 - y(14)) * kd - y(4) * Kr
  Delta = y(13)
  alpha = y(1) + y(3) / vel
  yd(1) = y(2)
  yd(2) = mua * alpha + muc * Delta - mup * y(10)
  yd(3) = -tom * y(1) - aom * alpha + com * Delta
  yd(4) = y(5)
  yd(5) = (thetadTOT - y(4)) * wf ^ 2 - 2 * Zf * wf * y(5)
  yd(6) = y(7)
  yd(7) = -comeq1 * Delta - tzob1 * y(7) - wbsq1 * y(6)
  yd(8) = y(9)
  yd(9) = -comeq2 * Delta - tzob2 * y(9) - wbsq2 * y(8)
  yd(10) = y(11)
  yd(11) = -fsd * Delta - tzofs * y(11) - wfsq * y(10)
  yd(12) = wact ^ 2 * (deltac - y(13)) - 2 * zact * wact * y(12)
  yd(13) = y(12)
  yd(14) = wkd * (thetaTOT - y(14))
Return
	data:
	  N = 4: y(1) = 0: y(2) = 1: y(3) = 0: y(4) = 0
	    yout = y(2)
	    timeh = 0: dt = 0.1: tstop = 8: istop = Int(tstop / dt)
	    timeX(1) = timeh: xy1(1) = yout
	    runm = Array(1, 0.5, 0.5, 1): jint = 1
	    m1 = 3: m2 = 2: k1 = 30: K2 = 20
	Return
	
	system:
	  yd(1) = y(3)
	  yd(2) = y(4)
	  yd(3) = (-(k1 + K2) * y(1) + K2 * y(2)) / m1
	  yd(4) = (K2 * y(1) - K2 * y(2)) / m2
	  yout = y(2)
	Return

Appendix F

Appendix F: Case 3 Chapter 15, Eigenvector and Lamda Matrices

eigenvectors (real)

0.0004	0.0124	0.0124	0.0005	0.0005	0	0	0	0	-0.0001	0	0	0	0
-0.0001	-0.0202	-0.0202	-0.0003	-0.0003	-0.0018	-0.0018	-0.0005	-0.0005	0.0037	-0.0013	-0.0013	0	0
1	0.117	0.117	-0.0234	-0.0234	-0.0251	-0.0251	-0.0067	-0.0067	0.0495	-0.017	-0.017	-0.0005	-0.0005
-0.0001	-0.0203	-0.0203	-0.0002	-0.0002	-0.0022	-0.0022	-0.0013	-0.0013	0.0019	-0.0002	-0.0002	0.011	0.011
0	-0.0264	-0.0264	-0.0142	-0.0142	-0.2348	-0.2348	-0.1563	-0.1563	-0.0657	0.0342	0.0342	-0.4143	-0.4143
0.0005	0.0077	0.0077	0.0007	0.0007	0.0486	0.0486	-0.0009	-0.0009	0.0037	-0.0009	-0.0009	0.0005	0.0005
-0.0001	-0.0335	-0.0335	-0.0053	-0.0053	-0.3091	-0.3091	0.0433	0.0433	-0.1315	0.0579	0.0579	0.005	0.005
0.0004	0.0069	0.0069	0.0006	0.0006	0.003	0.003	0.0262	0.0262	0.0057	-0.0018	-0.0018	0.001	0.001
-0.0001	-0.0295	-0.0295	-0.0045	-0.0045	-0.1153	-0.1153	-0.6423	-0.6423	-0.2006	0.1081	0.1081	0.0196	0.0196
0.0006	0.0068	0.0068	0.1664	0.1664	-0.0001	-0.0001	0	0	0.0004	-0.0001	-0.0001	0	0
-0.0001	-0.0446	-0.0446	0.5238	0.5238	0.0076	0.0076	0.002	0.002	-0.0142	0.0049	0.0049	0.0002	0.0002
0.0001	0.0622	0.0622	0.0093	0.0093	0.1464	0.1464	0.0914	0.0914	1	0.2142	0.2142	-0.8503	-0.8503
-0.0009	-0.0146	-0.0146	-0.0012	-0.0012	-0.0033	-0.0033	-0.002	-0.002	-0.0284	0.0094	0.0094	0.0127	0.0127
0.0005	0.0131	0.0131	0.0005	0.0005	0.0005	0.0005	0.0001	0.0001	0.0064	0	0	0	0

eigenvectors (imag)

0	0.0007	-0.0007	0.0001	-0.0001	0.0001	-0.0001	0	0	0	0	0	0	0
0	0.0255	-0.0255	0.0027	-0.0027	0.0007	-0.0007	0.0003	-0.0003	0	0.0002	-0.0002	-0.0009	0.0009
0	0.9931	-0.9931	0.0263	-0.0263	0.0102	-0.0102	0.0043	-0.0043	0	0.0028	-0.0028	-0.0127	0.0127
0	0.0267	-0.0267	0.0028	-0.0028	0.012	-0.012	0.0053	-0.0053	0	-0.0006	0.0006	0.0005	-0.0005
0	-0.0841	0.0841	-0.0012	0.0012	-0.0382	0.0382	-0.0371	0.0371	0	0.0132	-0.0132	0.644	-0.644
0	0.0102	-0.0102	0.001	-0.001	0.017	-0.017	-0.0015	0.0015	0	-0.0005	0.0005	-0.0004	0.0004
0	0.0011	-0.0011	0.0035	-0.0035	0.951	-0.951	-0.0267	0.0267	0	-0.0159	0.0159	0.0456	-0.0456
0	0.0089	-0.0089	0.0009	-0.0009	0.006	-0.006	0.0219	-0.0219	0	-0.0008	0.0008	-0.0009	0.0009
0	0.0012	-0.0012	0.003	-0.003	0.0619	-0.0619	0.7665	-0.7665	0	-0.0463	0.0463	0.0933	-0.0933
0	0.016	-0.016	-0.1037	0.1037	-0.0004	0.0004	-0.0001	0.0001	0	-0.0001	0.0001	0	0
0	-0.0096	0.0096	0.8519	-0.8519	-0.0031	0.0031	-0.0013	0.0013	0	-0.0008	0.0008	0.0037	-0.0037
0	-0.003	0.003	-0.0061	0.0061	-0.0684	0.0684	-0.0585	0.0585	0	0.9768	-0.9768	0.5262	-0.5262
0	-0.0187	0.0187	-0.0018	0.0018	-0.0076	0.0076	-0.0031	0.0031	0	-0.0138	0.0138	0.0067	-0.0067
0	0	0	0	0	0	0	0	0	0	0	0	0	0

lamda matrix (real)

-0.1346	0	0	0.0001	0.0001	-0.0001	-0.0001	0.0001	0.0001	0	0.0005	0.0005	-0.0174	-0.0174
0	-1.5168	0	-0.0001	0	0	0.0001	0	0	0	-0.0005	0.0001	0.0095	0.0067
0	0	-1.5168	0	-0.0001	0.0001	0	0	0	0	0.0001	-0.0005	0.0067	0.0095
0	0	0	-0.0315	0	0	0	0	0	0	0	0	-0.0006	-0.0008
0	0	0	0	-0.0315	0	0	0	0	0	0	0	-0.0008	-0.0006
0	0	0	0	0	0.4503	0	0	0	0	0.0001	0	-0.0012	-0.0006
0	0	0	0	0	0	0.4503	0	0	0	0	0.0001	-0.0006	-0.0012
0	0	0	0	0	0	0	-0.0641	0	0	-0.0001	0	0.001	0.0016
0	0	0	0	0	0	0	0	-0.0641	0	-0.0001	0	0.0016	0.001
0	0	0	0.0002	0.0002	-0.0002	-0.0002	0.0001	0.0001	-35.2156	0.0009	0.0009	-0.0314	-0.0314
0	0	0	0	0	0	0	0	0	0	-41.1262	0.0005	0.0041	-0.0009
0	0	0	0	0	0	0	0	0	0	0.0005	-41.1262	-0.0009	0.0041
0	0	0	0	0	0	0	0	0	0	0	0	-35.1016	0.0007
0	0	0	0	0	0	0	0	0	0	0	0	0.0007	-35.1016

lamda matrix (imag)

0	0	0	0	0	0	0	0	0	0	0.0006	-0.0006	-0.0028	0.0028
0	2.1407	0	0.0001	-0.0001	0	0	0	0	0	0	0.0005	-0.0073	-0.0099
0	0	-2.1407	-0.0001	0	0	0.0001	0	0	0	-0.0005	0	0.0099	0.0073
0	0	0	5.0992	0	0	0	0	0	0	0	0	-0.0008	-0.0006
0	0	0	0	-5.0992	0	0	0	0	0	0	0	0.0006	0.0008
0	0	0	0	0	19.4115	0	0	0	0	0	-0.0001	0.002	0.0023
0	0	0	0	0	0	-19.4115	0	0	0	0.0001	0	-0.0023	-0.002
0	0	0	0	0	0	0	29.2568	0	0	-0.0001	0	0.0022	0.0017
0	0	0	0	0	0	0	0	-29.2568	0	0	0.0001	-0.0017	-0.0022
0	0	0	0	0	0	0	0	0	0	0.0011	-0.0011	-0.005	0.005
0	0	0	0.0001	0.0001	-0.0001	-0.0001	0	0	0	43.4577	0.0005	-0.0155	-0.016
0	0	0	-0.0001	-0.0001	0.0001	0.0001	0	0	0	-0.0005	-43.4577	0.016	0.0155
0	0	0	0	0	0	0	0	0	0	0	0	60.2154	0.0003
0	0	0	0	0	0	0	0	0	0	0	0	-0.0003	-60.2154

Index: The VBA Language Elements

Statements	Common VBA Functions (Chapter 8)
Call, Chapter 10 Dim, Chapter 6 Do…Loop, 62 Do Until…Loop, 63 Do…Loop Until, 63 Do While…Loop, 62 Do…Loop While, 62 End, 72 End Function, 60 End If, 64 End Select, 65 End Sub, the last statement in all programs Exit Do, 62 Exit For, 61 Exit Function, 60 Exit Sub, 72 For…To…Next, 61 Function, Excel, Chapter 8 Function, VBA, Chapter 8 Function, User-defined, Chapter 8 GoSub…Return, Chapter 11 GoTo, Chapter 12 If…Then, 64 If…Then…End If, 64 If…Then…Else…End If, 64 If…Then…ElseIf…Then…Else…End If, 64 Option Base, Chapter 6 Option Explicit, Chapter 6 ReDim, Chapter 7 Select Case, 65 Sub, the first statement in all programs	Array, Chapter 6 Abs, 57 Atn, 57 Cos, 57 Exp, 57 Fix, 58 Int, 58 Log, 57 Round, 58 Sgn, 57 Sin, 57 Sqr, 57 Tan, 57
	Common Excel Functions (Chapter 8)
	Asin, 57 Acos, 57 Atan2, 57 Log10, 57 Log, 57 Max, 58 Min, 58 Minverse, 58 MMult, 58 MDeterm, 58 Transpose, 58 Round, 58 RoundUp, 58 RoundDown, 58
	Complex Excel Functions (Chapter 8)
	Complex, 59 ImReal, 59 Imaginary, 59 ImAbs, 59 ImArgument, 59 ImDiv, 59 ImProduct, 59 ImSum, 59 ImSub, 59 ImPower, 59 ImExp, 59

Input/Output
Cells(row, column), Chapter 4 Worksheets(" sheet*j* ").Cells(row, column), Chapter 4

Operators Used in Arithmetic and in Conditional Tests (Comparison/Logical)
Chapter 9, Section E

About the Author

Francis D. Hauser earned his PhD in electrical engineering from the University of Denver in 1972.

He is a dynamic-systems analyst in the motions and control of launch and reentry vehicles; multibody spacecraft; fixed- and rotor-wing aircraft; large, high-speed, oceangoing watercraft; landcraft; and wind-driven turbines.

He is a university and college lecturer in graduate, undergraduate, and continuing-education courses on general optimization theory; Newtonian mechanics (statics and dynamics); linear algebra; Fourier analysis; and conventional, modern, and space-vehicle control.

Made in the USA
Charleston, SC
17 June 2016